# Governance in the Extractive Industries

Greater understanding of the forms and consequences of investment and disinvestment in the extractive industries is required as a result of capitalist expansion, recent declines in global commodity prices, and claims that extractive sector projects, especially in the global south, are poverty reduction projects. This book explores emergent forms of governance in mining and extractive industry projects around the world.

Chapters examine efforts to govern extractive activities across multiple political scales, through intermediaries, instruments, technologies, discourses, and infrastructures. The contributions analyse how multiple micro-processes of rule reverberate through societies to shape the material conditions of everyday life but also politics, social relations, and subjectivities in extractive economies. Detailed case studies are included from Africa (Chad, Nigeria, Rwanda, and São Tomé and Príncipe), Latin America (Bolivia, Ecuador, and Peru), and the UN Climate Conference.

**Lori Leonard** is a Professor in the Department of Development Sociology at Cornell University, USA.

**Siba N. Grovogui** is a Professor in the Africana Studies and Research Center at Cornell University, USA.

# Routledge Studies of the Extractive Industries and Sustainable Development

# Governance in the Extractive Industries

Power, Cultural Politics and Regulation

Edited by
Lori Leonard and
Siba N. Grovogui

Routledge
Taylor & Francis Group

LONDON AND NEW YORK

from Routledge

First published 2017
by Routledge

2 Park Square, Milton Park, Abingdon, Oxfordshire OX14 4RN
52 Vanderbilt Avenue, New York, NY 10017

*Routledge is an imprint of the Taylor & Francis Group, an informa business*

First issued in paperback 2019

*British Library Cataloguing-in-Publication Data*
A catalogue record for this book is available from the British Library

*Library of Congress Cataloging-in-Publication Data*
A catalog record for this book has beeen requested

ISBN: 978-0-415-78688-1 (hbk)
ISBN: 978-0-367-35137-3 (pbk)

Typeset in Bembo
by Florence Production Ltd, Stoodleigh, Devon, UK

# Contents

# Contributors

**Omolade Adunbi** is Associate Professor in the Department of Afroamerican and African Studies (DAAS) and Faculty Associate in the Program in the Environment and in the Human Rights Program at the University of Michigan. He is a political anthropologist whose interests include natural resource extraction, governance, human and environmental rights, transnational institutions, multinational corporations, and the post-colonial state. He is the author of *Oil Wealth and Insurgency in Nigeria* (Indiana University Press, 2015). His current research focuses on the growing interest of China in Africa's natural resources and its interrelatedness with infrastructural projects. He teaches courses on transnationalism, globalization, violence, human and environmental rights, the post-colonial state, social theory, resource distribution and contemporary Africa, and culture and politics.

**Emily Billo** is Assistant Professor of Environmental Studies at Goucher College. She is a feminist geographer whose research focuses on resource governance and development in Ecuador. Her research examines the relationships between the Ecuadorian state, multinational oil companies, indigenous communities, and Corporate Social Responsibility programs in Ecuador's northern Amazon region. Her work has been published in *Geoforum*, *Progress in Human Geography*, and *Gender, Place & Culture*. Her current research investigates the criminalization of environmental protests, a violent strategy of repression carried out by the post-neoliberal Ecuadorian state as part of state-led mining operations.

**Brenda Chalfin** is Professor of Anthropology and Director of the Center for African Studies at the University of Florida. Bringing together cultural anthropology, geography, and political economy to establish new analytic points of entry to understanding political life in contemporary African states, her work addresses the complex functioning of national boundaries and frontiers, the popular production of infrastructure and urban public goods, non-territorial and maritime sovereignty, and the changing political valence of natural resource extraction in the context of late-capitalism. Her publications include *Shea Butter Republic: State Power, Global Markets, and the Making of an Indigenous Commodity* (Routledge, 2004), *Neoliberal Frontiers:*

*An Ethnography of Sovereignty in West Africa* (University of Chicago Press, 2010), and "Governing Off-shore Oil: Mapping Maritime Political Space in Ghana and the Western Gulf of Guinea," *South Atlantic Quarterly*, 2015.

**Kristin Doughty** is Assistant Professor of Anthropology at University of Rochester. She is the author of *Remediation in Rwanda: Grassroots Legal Forums* (University of Pennsylvania Press, 2016) that examines how Rwandans navigated the combination of harmony and punishment in local courts purportedly designed to rebuild the social fabric in the wake of the 1994 genocide. She is currently conducting ethnographic research on the cultural politics of energy and unity in post-genocide Rwanda, with a focus on methane extraction in Lake Kivu.

**Siba N'Zatioula Grovogui** is Professor of Africana Studies at Cornell University and was previously Professor of International Relations Theory and Law at The Johns Hopkins University. He is the author of *Sovereigns, Quasi-Sovereigns, and Africans: Race and Self-determination in International Law* (University of Minnesota Press, 1996) and *Beyond Eurocentrism and Anarchy: Memories of International Institutions and Order* (Palgrave, 2006). He is from Guinea, where he attended law school before serving as law clerk, judge, and legal counsel for the National Commission on Trade, Agreements, and Protocols. He is currently working on two manuscripts: *Otherwise Human: Humanitarian Discourses and their Subjects* and *Future Anterior: The International, Past and Present*.

**Lori Leonard** is Associate Professor in the Department of Development Sociology at Cornell University and Director of the Polson Institute for Global Development. She is the author of *Life in the Time of Oil: A Pipeline and Poverty in Chad* (Indiana University Press, 2016), which is based on 12 years of ethnographic fieldwork around an oil pipeline project in Chad that the World Bank described as a "model" extractive industry project. Her research interests span medical anthropology, environmental studies, and gender studies. She teaches courses on gender and global economic integration, the sociology of waste, and ethnographic fieldwork.

**Fabiana Li** is Associate Professor of Anthropology at the University of Manitoba. She is the author of *Unearthing Conflict: Corporate Mining, Activism, and Expertise in Peru* (Duke University Press, 2015), that is based primarily on fieldwork in communities neighboring Peru's largest gold mine. Her work has also been published in edited collections on Corporate Social Responsibility and the anthropology of water. She teaches course on cultural anthropology, social theory, globalization, and the environment.

**Tom Perreault** is Professor of Geography at Syracuse University. His research and teaching focus is on political ecology and environmental justice. His work examines resource extraction, water governance, rural livelihoods, indigenous/campesino social movements, and agrarian political economy in the Andean region. He is author of over 50 journal articles and book chapters; editor of *Minería, Agua y Justicia Social: Experiencias Comparativas de Perú y Bolivia* (Cusco: Centro Bartolomé de las Casas. and La Paz: PIEB, 2014); and lead editor (with Gavin Bridge and James McCarthy) of the *Routledge Handbook of Political Ecology* (London: Routledge, 2015). He has been awarded three Fulbright grants, and has been visiting scholar at the Universidad de los Andes (Bogotá, Colombia), and Universidad Católica del Norte (San Pedro de Atacama, Chile).

**Simone Pulver** is Associate Professor of Environmental Studies and Director of the Interdepartmental PhD Emphasis in Environment and Society at the University of California, Santa Barbara. Her research focuses on the intersection of economic action and environmental harm and seeks to integrate theoretical frameworks relating to global governance and economic and environmental sociology. She was previously the Joukowsky Assistant Research Professor at Brown University's Watson Institute for International Studies with a joint appointment in the Center for Environmental Studies.

**Gisa Weszkalnys** is Assistant Professor in Anthropology at the London School of Economics. She is the author of *Berlin, Alexanderplatz: Transforming Place in a Unified Germany* (Berghahn, 2010) and co-editor of *Elusive Promises: Planning in the Contemporary World* (Berghahn, 2013). Her current research explores speculations, expectations, and fears regarding future oil extraction in São Tomé and Príncipe, an emergent oil economy in the Gulf of Guinea.

# Acknowledgments

We would like to thank the Polson Institute for Global Development, the Africana Studies and Research Center, the Atkinson Center for a Sustainable Future, and the Institute for African Development, all at Cornell University, for support for the workshop that was the genesis of this volume. Lori Leonard would also like to thank the Rockefeller Foundation, Bellagio Center, for the academic writing fellowship that provided concentrated time and an unparalleled environment in which to work on this project.

# Introduction

## Governing in the extractive industries

*Lori Leonard and Siba N. Grovogui*

A book on governing in the extractive industries may seem counter-intuitive in the current environment. Commodity prices are in a slump, and are expected to remain low into the coming decade (Avent 2013). Yet, the present effort is motivated, at least in part, by the observation that the extractive industries will remain vital to the world's energy infrastructure for the foreseeable future, especially since alternative energy sources depend, paradoxically, on them. Moreover, during the commodities boom that preceded the current slump, many countries in the global south and in the global north came to depend on investments in extractives for jobs, revenue, technology and infrastructure. Beyond their continuing economic importance, extractive projects and infrastructures have become entangled in the social, cultural, and political lives of their hosts (Barrett and Worden 2014). Extractive projects have acquired new rationales in the post-development era, and as they have come to be framed as poverty reduction projects they have incorporated new sets of actors, logics, and modes of operation (World Bank 2012). All of these trends are important to track.

The Paris Agreement, the outcome of the 2016 Paris Climate Conference, stresses the need for an energy transition to meet climate change objectives. Yet, nearly all energy forecasts suggest that fossil fuels will continue to provide the bulk of the world's energy needs in the coming years. The International Energy Agency (IEA) predicts that demand for energy will increase by as much as 30 percent by 2040, as the world's population expands by more than one billion and per capita energy consumption rises as more people enter the middle classes (IEA 2016). Investments in renewable energy are growing at a faster pace than investments in non-renewables, but the transition to low-carbon energy sources is still expected to take decades. A less discussed, but still important, aspect of the transition is the extent to which renewable energy sources depend on extractive activities. After all, solar panels are built with aluminum frames and nuclear energy is produced from the fissioning of uranium atoms and therefore depends on the extraction of uranium ore. A second line of response to climate change is making carbon-based energy cleaner, safer, and more efficient, and while these efforts might change the nature of extractive activities they will not eliminate the need for them.

The demand for energy and in-the-ground resources has long been characterized by booms and busts. The latest surge in investment began in the late 1990s and lasted through the first decade of the current century. This boom, which economists have described as a "commodities super-cycle" (Avent 2013), is said to have been the longest and largest since 1900 (World Bank 2009) and to have been spurred by rapid industrialization and urbanization in China that raised concerns about global commodity shortages. Much like the information technology (IT) bubble that led to what Thomas Friedman (2005) called the "massive overinvestment of billions of dollars in fiber-optic telecommunications cable" that, seemingly overnight, connected the world and made the reorganization and (further) globalization of business possible, the latest commodities super-cycle transformed national and regional economies. The World Bank's investments in oil, coal, and gas rose four-fold during this period and its energy investments rose three-fold in the poorest countries in the world, where governments benefitted from the bonanza but at the same time developed dependencies on natural resources (Redman et al. 2015).

The nature and scope of these resource dependencies vary across regions and countries. Foreign direct investment in extractives is but one measure of this variation. According to the United Nations Conference on Trade and Development's (UNCTAD) 2015 *World Investment Report*, investments in the extractive sector in Sub-Saharan Africa accounted for nearly one-third of all foreign direct investment in that region (UNCTAD 2015). That proportion was similar to previous years and was not declining, even after commodity prices had fallen from their peak. By contrast, the extractive sectors in Latin America and the Caribbean accounted for a much smaller proportion of overall foreign investment and these investments were sensitive to the downturn in commodity prices. Yet these comparisons yield a picture that is incomplete, since hydrocarbons in Latin America are produced largely by state-owned companies. Indeed, this "resource nationalism" (Kohl and Farthing 2012; Zimmerer 2015) represents another key point of contrast with Sub-Saharan Africa.

Differences in the relative primacy of primary commodities in national and regional economies and in the organization of the extractive sectors matter. While not determinative, they structure risks and opportunities. Resource dependence may mean that oil, gas, and mining companies are given favorable treatment, reflected in state-guaranteed contractual privileges and immunities, or that the state's capacity to negotiate with financiers and companies is impacted (UNCTAD 2015). Some states or regions are more susceptible than others to the idiosyncrasies of international financial institutions and large transnational firms. Not only is Sub-Saharan Africa (and there is a considerable degree of internal differentiation on the continent as well) in a weak negotiating position relative to investors, African states also invite more external mandates, ranging from the minutiae of revenue allocation to the conduct of state–civil society relations, and burdensome conditionalities.

Beyond the commodities super-cycle, the broad context for the chapters in this book is the re-orientation of national policies subsequent to the collapse of the Communist system and the demise of the developmental state. The retreat of the state and the ascendency of capital, particularly finance capital, to preeminence has had profound effects in terms of structuring the collective imaginaries of the possible and the language of politics itself. This is evident in the rise of global regimes of governance and companies as development actors. These turns of events are formulated as responses to the retreat of the state. It is in this context that extractive industry projects, especially in the global south, have come to be framed in terms of human development or poverty reduction objectives. In a speech to the Asia Society, Rex Tillerson, now the US Secretary of State but then the CEO of ExxonMobil, the world's largest private oil company, described providing affordable fossil fuels to people in the global south as a "humanitarian imperative." According to Tillerson, "the need to expand energy supplies has a humanitarian dimension that should inform and should guide our energy policy dialogue" (Tillerson 2013).

One consequence of this positioning of extractive activities as a form of humanitarianism has been to pit the needs of the poor against efforts to mitigate climate change through reductions in the use of fossil fuels (Klare 2014). When Tillerson was asked about global warming and the extent to which mitigating it was a priority as part of a recorded conversation at the Council on Foreign Relations, he said, "there are much more pressing priorities," and went on to describe the "billions of people living in abject poverty around the world" and their needs for electricity and "fuel to cook their food on that's not animal dung" (Council on Foreign Relations 2012). Thus, poverty reduction and humanitarianism have become key tropes in the operation of the extractive sector—a sector that earlier eschewed the notion that its operations were in any way entangled with the fates of local communities and that organized its operations in ways that highlighted this separation and detachment (Ferguson 2005). The chapters in this volume pick up on multiple formulations of extraction as promise and point to the centrality of energy infrastructures in shaping plural and often competing visions of the future and the good life.

Another consequence of the positioning of extractive industries within the humanitarian frame is to shift the focus of analysis away from the role of the state or at least to broaden that focus to include global forms of governance, global ethics regimes, and corporate and transnational actors. Extractive industry projects incorporate global standards, voluntary codes of conduct, and corporate social responsibility schemes that are enacted with the help of a profusion of actors, instruments, and technologies. States are not absent from this picture, and in some instances are engaging with extractive projects in novel ways. The "new humanitarianism" includes, for example, government-funded aid and development agencies working with mining companies to deliver job-skills training programs (York 2012). At the same time, other actors resist the formal logics of states and companies or they replicate and redeploy them in novel ways. The chapters in this book look at particular assemblages of state, corporate, and transnational forms and at emergent energy infrastructures and their consequences within and beyond the spaces of their enactment.

## The project

This is a book about governing and governance in the extractive industries. It is a book about the ways extractive industry projects are organized, articulated, and carried out under late-neoliberalism and of the implications for people, the environment, and society. We are mindful that the terms "governing" and "governance" are over-used to the point of being hackneyed. As Claude Offe said, "When one refers to something as an instance of 'governance,' one has not yet expressed much—exactly because of the multitude of possible meanings" (Offe 2009, 554). For this reason, he refers to these terms as "empty signifiers" and describes them as "merely verbal frame[s] for largely exchangeable contents" (Offe 2009, 561). One way to manage that "multitude of possible meanings" would be to tell readers how we intend to use the terms or to restrict contributors' use of them to particular disciplinary or theoretical lineages. We take a different tack.

We propose, instead, to exploit the indeterminacy of these terms with the view that there can be something productive about this uncertainty. Working with empty categories or signifiers makes it possible to incorporate different people, ideas, or elements. Ernesto Laclau (2007) used this logic to write about populism and the political importance of empty signifiers for movement politics, and others have taken up his reasoning to describe how the Occupy Movement, for example, functions as an umbrella concept to link different kinds of demands and different groups of people together and to organize them against a common "enemy"—in this case, the 1 percent (Savage 2014). Precisely because Occupy is an empty category, it can be used to organize, assemble, and arrange different groups and provide them with a political identity.

There are, of course, clear differences between this example and our own initiative, and the parallels with our project are partial at best. We have no common "enemy," and to the extent that we have demands, they take the form of issues to which we wish to draw readers' attention. Still, what we recognize as useful from this formulation is the potential empty signifiers or categories afford to form new alliances and to attract and organize people who might not otherwise be found together. Most edited volumes on the extractive industries focus on a specific commodity, industry, or world region (see for example Appel, Mason, and Watts 2015; Barrett and Worden 2013; Behrends, Reyna and Schlee, 2013; Veltmeyer and Petras 2014). There are good reasons for this. At the same time, we wanted to see what conversations and openings might emerge from the effort to draw scholars together from across disciplines and theoretical traditions who work on different industries in different regions of the world.

Most of the chapters that appear in this book were produced for a workshop held at Cornell University in October of 2015, though some were solicited from scholars who were unable to participate in that original workshop. Participants were asked to contribute chapters with little more direction than the workshop title—"Governing Extraction"—and a brief statement of our interest in novel forms of governance and their implications in and beyond sites of extraction. Participants were intentionally drawn from different disciplines: geography, political science, anthropology, sociology, and the law, and interdisciplinary fields such as development studies, environmental studies, and regional studies programs. They were also selected because of their work in and on a range of industries and sites: Peru, Bolivia, Ecuador, Rwanda, Chad, Nigeria, São Tomé and Príncipe, and the UN Climate Conference. Scholars working in Latin America and Sub-Saharan Africa are over-represented in this volume because these seem to be the regions of most ferment and experimentation in extractive governance. Our sample is therefore not representative in any sense; it is experimental and even haphazard.

In the final chapter and post-script to this book, Brenda Chalfin draws out several critical threads from this experiment and the resulting collection of chapters that illustrate important dimensions of the character of extractive governance today. These threads cut across and provocatively realign the themes we elaborate below and use to regroup the chapters.

## The book

The chapters in this book are organized into three sections to reflect important dimensions of governing in the extractive industries addressed by the authors. Chapters in the first section of the book, by Tom Perreault, Emily Billo, and Siba N. Grovogui, explore the legal, socio-political, and institutional contexts of extraction through textual and socio-political analyses of legislation. While modern governance is not reducible to the enactment of laws, contracts, and decrees, these instruments remain critical reference points for state and corporate activities in the extractive industries. The chapters in the first section of the book highlight the indeterminacies of law and how it is proliferates power through the trope of the rule of law, the legitimation of force, and even through the recognition of multiple and competing interests around extraction.

Tom Perreault examines the legal infrastructure governing extraction in Bolivia as a case in which the proliferation of laws designed to appease different constituencies seems to both reflect and exacerbate tensions and contradictions that remain to be worked out in practice. His chapter, "Tendencies in Tension: Resource Governance and Social Contradictions in Contemporary Bolivia," examines a trio of recent laws taken up by the Bolivian state and contemplates what their co-existence might mean for the politics of extraction. Mining laws that incentivize corporate investment by providing stable structures for consultation of affected communities with the intent of eliminating conflict stand in awkward relation to laws emanating from popular movements that instantiate a form of environmental awareness, framed as an "indigenous cosmovision," and privilege living "in harmony and equilibrium with Mother Earth."

In the Bolivian context, the formation of political opposition to corporate interests and state subordination to those interests was intended not only to communicate displeasure but also to transform Bolivian politics. The most spectacular outcome of that opposition movement so far has been the rise of Evo Morales to the presidency—making him the first chief executive of indigenous descent. Yet, Perreault's examination of the legal armature of Bolivia begs the question of how it will be possible for the Morales government to reconcile these competing and seemingly contradictory orientations moving forward. Perrault invites a reading of these laws that moves beyond the interests of industry and those of "well-established social groups" to contemplate their possible outcomes over time. He does not offer an empirical analysis of the implementation of law so much as expose the juridical underpinnings of resource conflicts yet to come.

Emily Billo takes readers to Ecuador to examine the policing practices, legal proceedings, institutional arrangements, and social conditions "that extend . . . and permit practices of criminalization in the mining industry." In her chapter, "Mining, Criminalization, and the Right to Protest: Everyday Constructions of the Post-Neoliberal Ecuadorian State," Billo constructs her arguments against the backdrop of two tenets of the ideological predicates of the government of President Rafael Correa: an emphasis on the primacy of social rights and an experiment in so-called post-neoliberalism. Billo notes that Ecuador's post-neoliberal governing model is still firmly rooted in resource extraction and that this presents a problem for the actualization of social rights. The government's approach to the resolution of the tension between its own interest in steady income from mining and the interests of environmental groups is social control. This is accomplished through the enactment of laws and decrees that criminalize many aspects of social and political activities by non-governmental groups that would undermine the stability of mining.

The apparent tension between the Correa government and activist groups over extraction suggests that civil society groups have developed interests in the extractive industries that range from insistence on participation and consultation in developing the legal framework of extraction to shaping the terms of the distribution of oil revenues. These demands present the Correa government with a quandary (one that we might imagine could also confront the Morales government): ignore them and resort to authoritarian repression or agree to them and risk prying eyes and possible revelations about the co-dependencies and machinations that bind governments and extractive industries to one another.

In the final chapter in this section, Siba N. Grovogui takes aim at the legal superstructure erected around the Chad Pipeline Project under the rubric of the rule of law. In "Preserving Illusions: The Rule of Law and Legitimacy under the Chad Pipeline Project," Grovogui explores the ideological yet strategically instrumental uses of the idiom of the rule of law. Unlike the architects of the Chad experiment, who deploy the ideology of the rule of law as a positively neutral and historically unquestionable bedrock of development, Grovogui focuses on the utility of the rule of law as a metaphor for state, corporate, and global governance. He points to the importance of language, and specifically to the overreliance on the idiom of the rule of law, in framing discourses of equity, justice, and transparency while showing how this language justifies institutional arrangements that are problematic for those constituted internally in discourse as requiring protection.

Grovogui cautions that the production, reproduction, and circulation of ideas and institutions associated with the rule of law may amount to an artful deflection of attention from the nuts and bolts of governance. It is a truism that laws, administrative edicts, and contractual agreements between states and corporations are grounded in social contexts and structures of interpretation and execution, but Grovogui highlights some of these circumstances that give meaning to the rule of law in Chad. He argues that the rule of law has multiple valences whose effects depend on the constitutional order in which government and the governed are bound by and subject to the law, and that terms and tropes associated with the rule of law may be specific to a context but that subjects do not take them as self-evidently applicable to their circumstances. Contestation therefore emerges as a supplement to governance and one of its singular characteristics. Taken in this context, he argues that the injection of the rule of law in public discourse as justification for extraction should be viewed as the introduction of non-consensual and negotiated paradigms and axioms of society, social relations, and governance.

The chapters in the second section of the book, by Omolade Adunbi, Kristin Doughty, and Simone Pulver examine different kinds of claims and counter-claims to natural resources or to the management of them. The chapters all highlight contests over the benefits of extractive activity and trace the ways local and transnational framing devices—ranging from claims of antecedent rights to global warming, explosive lakes, and national unity—are deployed to mobilize actors, legitimize specific claims, and shape the means and ends of extractive activity. The chapters also reveal how fragile or ephemeral these accomplishments are and the constant work that is required to hold hard-won ground or to maintain coalitions, support, or moral authority.

Omolade Adunbi's chapter, "We Own This Oil: Artisanal Refineries, Extractive Industries and the Politics of Oil in Nigeria," focuses on contestations between the Nigerian state and groups of Delta "youth"—a category used to refer to the un- and under-employed in Nigeria's oil economy more than to those of a certain age—who are behind the construction and operation of an impressive array of artisanal refineries. Rejecting the language of criminality used by the state to refer to these activities, Adunbi explores the drama animating claims and counter-claims over who owns the oil, how extraction should be carried out, and how the spoils should be distributed. Delta youth have moved beyond protest to actualize sovereign claims to "their" natural resources by building infrastructure that allows them to refine crude oil and sell it on the global market. Their actions not only challenge the political authority of the state but also raise interesting analytical questions about how to view the equally moralizing language of those presumed by the state to be saboteurs and thieves. According to Adunbi, the youth who run the refineries claim prior ownership of the land and have constructed a narrative of precedence or ascendant privilege from this principle of sovereign antecedence over claims by the state and oil corporations and have extended the principle of ascendency to a right to construct their own infrastructure for refining and distributing oil in defiance of state laws and contractual agreements.

Adunbi's goal is to rethink conflicts over oil in the Delta by "looking critically at the emergent oil infrastructure constructed by youth as competing with state oil infrastructures in the production of crude oil." He insists on examining the actions of the youth who operate the refineries on their own terms, and therefore on seeing their practices as constituting an alternative to state-led extractive development. In his ethnographic account of artisanal production, he shows how local operators assemble skills and technologies acquired from producing gin, a practice that dates to the colonial period and prohibition and has been passed down in families, as well as from work as pipefitters and welders for transnational oil companies to develop a sophisticated production system that promises benefits to local families and therefore garners communal support. The Delta emerges as a site of contestation and opportunity for youth to successfully compete with the state in setting up alternative channels of production and distribution of crude oil and, at least for now, to respond to the pleas of residents of the creeks who assert that their interests are harmed by the relationship between the state and corporate conglomerates.

Kristin Doughty expands on the connection between the social and political context of extraction and the imagination of the political economy in her chapter on "Converting Threats to Power: Methane Extraction in Lake Kivu, Rwanda." Doughty argues that in post-genocide Rwanda the politics of energy and national unity are deeply interconnected. The setup for her arguments is a project by an American company to extract methane from Lake Kivu—a lake that has been described as "dangerous" and "a freshwater time bomb" following a deadly eruption in a similar lake in Cameroon in the 1980s. In post-conflict Rwanda, the risks of methane release from the lake are woven into discussions of survival and social reproduction. Efforts to promote the methane extraction project and to provide a sustainable source of electrical power to the "new," post-genocide Rwanda are powered by claims about how these extractive activities will also reduce the dangerous levels of unstable gases dissolved in the lake, thus converting a threat into a benefit.

Yet, as Doughty shows, the decision to convert a potential agent of death— methane gas—into life-affirming opportunities through the provision of electrical power requires the alliance of an eclectic network of actors. She also shows the considerable work that is required to sustain this alliance through the case of a sudden and inexplicable color change in the lake that threatened to unravel the work of the conversion narrative. Doughty argues that the discovery of a source of energy in Lake Kivu, precisely because it implied the conversion of agents of death into a source of life, gave further impetus to domestic policies of national unity and reconciliation. In making this argument, she finds utility in Dominic Boyer's (2014) concept of "energopower" in which the management of life and populations "depends in crucial respects upon . . . the harnessing of electricity and fuel and vice-versa." This analytic frame, as she says, sheds "light on the imbrications of the cultural politics of energy and unity which are important to understanding socio-political dynamics across the African continent and elsewhere."

Simone Pulver approaches questions about the management of natural resources through an impressive re-articulation of the Habermasian notions of participation and the public sphere to explain the apparent efficacy of non-governmental organizations at the UN Climate Conference. "A politics of the public sphere: ENGOs and oil companies in the international climate negotiations, 1987–2001," allows Pulver to bring to light the successful interventions of environmental groups to that event. Her chapter does not conflate the domestic constitutional sphere with the global one. What commands her attention is how the less regulated and more chaotic mechanisms of global governance engender opportunities to consider multiple forms of constitutionalism. She shows how the very porosity of the global sphere allows for continuous constitutional innovations or, alternatively, permanent opportunities for non-governmental groups to define and redefine the terms and boundaries of the permissible with regard to policy. Pulver does not deny that the rules and procedures governing access and participation at the UN Climate Conference were conceived by corporate-friendly governments, but she demonstrates persuasively that the rules and procedures that were instituted created the conditions for imaginative environmental NGOS (ENGOs) to define the terms under which they engaged their state and corporate counterparts during the negotiations. The ENGOs' tactics, which included publicity stunts and shaming, proved in some cases to deter states and corporations from undermining the very purpose of the conference and allowed activists to turn the space of protest granted to them under legal arrangements and consultative processes into effective instruments of governance.

While focusing on the shifting political influence of oil companies and ENGOs in international climate negotiations, Pulver points to the emergence of political arenas in which legitimacy is predicated on robust exchanges with unpredictable outcomes within emergent spheres of deliberation whose forms evolve based on political opportunities and the need to confront apparent challenges. In this Habermasian model of deliberative democracy, opportunity is born out of two events. The first is the insistence of the environmental community that it be taken seriously and therefore included in matters in which its members have expertise, including discussions of global climate change. The other is the softening of the intransigent positions of oil companies in light of evidence of climate change and the rising global demand for action. According to Pulver, in the subsequent international conference on the climate a triangular relationship emerged between states, industry, and ENGOs in the form of a public sphere: a space of deliberation not entirely owned by a single participant and whose content substantively was the product of all participants.

The third and final section of the book engages a series of questions about the roles of experts, expertise, and information in extractive projects. The

chapters by Gisa Weszkalnys, Fabiana Li, and Lori Leonard highlight a paradox
for governance in that the democratic energy unleashed by the fall of Com-
munism and the proliferation of information technologies has run up against
new state and corporate practices to control knowledge and information
and thereby maintain leverage in decision-making. These chapters illustrate
the expanding role of various kinds of experts and expertise in the extractive
sector and show how economic, biomedical, and environmental information
is deployed by experts or filtered through them to shape stories and experiences
before, during, and after extractive projects.

"Preventing the Resource Curse: Ethnographic Notes on an Economic
Experiment" is the title of Gisa Weszkalnys's account of the role economic
experts played in the island state of São Tomé and Príncipe ahead of extractive
activity. Whereas initiatives by international financial institutions in the
engineering of social experiments tied to extraction are common, Weszkalnys
presents a case in which a country intent on a prophylactic "cure" for the resource
curse invited the aid of an expert, the US economist Jeffrey Sachs, to help
transform the island's economy. The concern of the official requesting the
assistance, President Fradique de Menezes, was to rescue his country from the
turmoil associated with the prospect of oil. Sachs and the retired US State
Department official who conveyed the message to him advised the island state
about ways in which to avoid the fate of other states in what they saw as the
"'bad' neighborhood of the Gulf of Guinea." There, other ill-equipped
countries had gone it alone against giant oil corporations only to be disappointed
by the outcome of the deals they had made.

Weszkalnys's chapter charts the course of what becomes of an appeal to a
well-connected expert to provide advice and deploy his connections to render
the opaque world of the oil industry and international agreements legible to
the officials of a small country. The chapter offers insight into the manner in
which expertise is translated into policy or, in this particular case, how experts
like Sachs translate economic theory into actual practice; how they formulate
responses to ideas like the resource curse; what they consider proper economic
behavior; and what kinds of relations they imagine between the state and its
citizens, on the one hand, and economic regimes and the social compact, on
the other. Weszkalnys uses the case of São Tomé and Príncipe to illustrate the
contribution of economics to the constitution of the imaginary of the political economy. The chapter is a response to the question: What happens when
economic theories are enacted in specific contexts? The answer is that the details
are messier than any technical advice would suggest. As she says, "economic
tools and theories are not simply imposed or implemented in different contexts but rather collide with preexisting conditions and result in complex
articulations."

Fabiana Li adds new dimensions to our understanding of governing extraction at the micro-level in her chapter, "Illness, Compensation, and Claims for
Justice: Lessons from the Choropampa Mercury Spill." In contrast to Pulver's
global forum, the setting for Li's chapter is Choropampa, a small Peruvian town

that was the scene of an accident involving a truck that spilled the liquid mercury it was carrying from the Yanacocha gold mine in northern Peru to the capital city of Lima. Li chronicles the struggles of residents of Choropampa to receive redress for their injuries and the contamination of their environment. In doing so, she shows that emergent forms of environmental governance are not always deliberative and consensual, at least in the manner described by Habermas and Pulver. Minera Yanacocha, the mine operator, assumed some responsibility for clean-up and payment of damages, but it understood its responsibilities in terms that were very different from the expectations of the affected communities for fair and just treatment.

In her account of the accident as an effect of the rapid expansion of large-scale mining in Peru, Li focuses on the conflicts surrounding the toxic spill to interrogate the nature of risk in the context of neoliberal reforms and the expansion of the extractive sector in Peru. Li suggests that the Choropampa case offers a stark example of how the lives of people who lived well outside the area described by Minera Yanacocha as its direct "area of influence" are shaped by extractive activities and expert assessments of risk and harm. These effects included the physical and environmental sequelae of mercury poisoning but also biomedical subjectification, as residents struggled to turn themselves into "legitimate" claimants—a process that led to infighting and jealousy, dividing communities and turning families "on the side of the road" against each other as they fought over the uneven and seemingly arbitrary allocation of payments. The analytically significant fact in Li's narrative is that Minera Yanacocha considers the Choropampa incident a closed case while the disaster lives on in the experiences of residents and continues long after the fact to structure subjectivities and everyday interactions.

Lori Leonard's chapter, "Wars of Words: Experts, Oil, and Environmental Governance in Chad," deals with the growing emphasis in the extractive sector on transparency and accountability. Her particular interest is in the role played by the International Advisory Group (IAG), a group of expert monitors hired by the World Bank to track and report on the progress of an oil pipe-line project in Chad that was supposed to double as a poverty reduction project. The IAG was the collective analogue to Jeffrey Sachs in São Tomé and Príncipe, though with one crucial difference: it was expected to act as a third-party mediator to help establish the facticity of competing claims for the purpose of their adjudication. She argues that calls for expert monitors to be "the new transparency powerbrokers" (Mol 2010) and to exert normative pressure on oil companies and their international financial backers to align their activities with poverty reduction goals are based on faulty assumptions about information disclosures and how they work.

Leonard's analysis moves in two directions. Drawing on the work of Arthur Mol, she points to the immense differences in the relative capacities of actors in extractive industry projects—and especially in the Chad project—to collect, manage, and disseminate information to argue that informational resources are always already the products of power-laden processes. Her analysis also demonstrates the corollary, which is that experts are not outside informational flows but are instead implicated in different ways in them. In drawing on a second body of scholarship that highlights the performative aspects of transparency, Leonard shows how, in the Chad project, performances of transparency—which mostly took the form of report writing—had the effect of drawing the public's attention to the standards regime and to the oil companies' compliance with it, even as those companies were redefining the terms of their agreements. Her analysis problematizes efforts by organizations such as Oxfam America to advocate "watching the watchdogs" and to expand the layers of monitoring and normalize the use of expert monitoring bodies in high-risk and high-profile projects (Oxfam America 2011).

In the final chapter in this volume, which also serves as a post-script, Brenda Chalfin reads across these chapters in an effort to draw out what is new or noteworthy about extractive governance in the early twenty-first century. In particular, she draws attention to the complex, unsettled, uncertain, and increasingly "lateral" political terrain on which natural resource extraction is taking place. She does this to argue that more and more people have a stake—or some kind of stake—in the future of extraction and, at the same time, and not unrelatedly, that the governmental apparatus around extraction is ever-expanding in its size, scope, and import. While emergent governing frameworks, as Chalfin says, are "difficult to fully decipher," she begins, with her chapter, to trace some of the contours of the future of extraction and extractive futures.

## References

Appel, Hannah, Arthur Mason, and Michael Watts, eds. 2015. *Subterranean Estates: Lifeworlds of Oil and Gas*. Ithaca: Cornell University Press.

Avent, Ryan. 2013. "Commodity Prices: Rocks for the Long Run." *The Economist*, June 12. www.economist.com/blogs/freeexchange/2013/06/commodity-prices.

Barrett, Ross, and Daniel Worden, eds. 2014. *Oil Culture*. Saint Paul: University of Minnesota Press.

Behrends, Andrea, Stephen P. Reyna, and Günter Schlee, eds. 2013. *Crude Domination: An Anthropology of Oil*. New York: Berghahn Books.

Boyer, Dominic. 2014. "Energopower and Biopower in Transition." Special Collection. *Anthropological Quarterly* 87 (2): 309–333. doi: 10.1353/anq.2014.0020.

Council on Foreign Relations. 2012. "CEO Speaker Series: A Conversation with Rex Tillerson." June 27. www.cfr.org/world/ceo-speaker-series-conversation-rex-w-tillerson/p35286.

Ferguson, James. 2005. "Seeing Like an Oil Company: Space, Security, and Global Capital in Neoliberal Africa." *American Anthropologist* 107 (3): 377–382.

Friedman, Thomas. 2005. "It's a Flat World After All." *New York Times Magazine*, April 3. www.nytimes.com/2005/04/03/magazine/its-a-flat-world-after-all.html.

International Energy Agency. 2016. *World Energy Outlook 2016*. Paris: OECD/IEA.

Klare, Michael. 2014. "Big Energy is Just Like Big Tobacco—and That's Terrible News for Everyone." *Salon*, May 28. www.salon.com/2014/05/28/big_energy_is_just_like_big_tobacco_and_thats_terrible_news_for_everyone_partner/.

Kohl, Benjamin, and Linda Farthing. 2012. "Material Constraints to Popular Imaginaries: The Extractive Economy and Resource Nationalism in Bolivia." *Political Geography* 31 (4): 225–235.

Laclau, Ernesto. 2007. *On Populist Reason*. London: Verso.

Mol, Arthur P.J. 2010. "The Future of Transparency: Power, Pitfalls and Promises." *Global Environmental Politics 10* (3): 132–143.

Offe, Claude. 2009. "Governance: An Empty Signifier?" *Constellations 16* (4): 550–562.

Oxfam America. 2011. *Watching the Watchdogs: Evaluating Independent Expert Panels that Monitor Large-scale Oil and Gas Pipeline Projects*. Washington, DC: Oxfam.

Redman, Janet, Alexis Durand, Maria C. Bustos, Jeff Baum, and Timmons Roberts. 2015. *Walking the Talk? World Bank Energy-Related Policies and Financing 2000–2004 to 2010–2014*. Washington, DC: Institute for Policy Studies.

Savage, Ritchie. 2014. "Reflections on Laclau." *OccupyWallStreet*, April 20. http://occupywallst.org/article/reflections-laclau/.

Tillerson, Rex. 2013. "A Business Perspective on Global Energy Markets and Asia." June 13. http://corporate.exxonmobil.com/en/company/news-and-updates/speeches/a-business-perspective-on-global-energy-markets.

United Nations Conference on Trade and Development (UNCTAD). 2015. *World Investment Report 2015—Reforming International Investment Governance*. New York and Geneva: United Nations Conference on Trade and Development. http://unctad.org/en/PublicationsLibrary/wir2015_en.pdf.

Veltmeyer, Henry, and James Petras, eds. 2014. *The New Extractivism: A Post-Neoliberal Development Model or Imperialism of the 21st Century?* London: Zed Books.

World Bank. 2009. *Global Economic Prospects: Commodities at the Crossroads*. Washington, DC: The World Bank. http://siteresources.worldbank.org/INTGEP2009/Resources/10363_WebPDF-w47.pdf.

World Bank. 2012. *The World Bank Group in Extractive Industries: 2012 Annual Review*. Washington, DC: The World Bank Group. http://documents.worldbank.org/curated/en/165421468326682912/pdf/778660AR0WBG0E00Box377313B00PUBLIC0.pdf.

York, Geoffrey. 2012. "In West Africa, a Canadian Mining Company Pioneers 'the New Humanitarianism.'" *The Globe and Mail*, March 20. www.theglobeandmail.com/news/world/in-west-africa-a-canadian-mining-company-pioneers-the-new-humanitarianism/article535009/.

Zimmerer, Karl. 2015. "Environmental Governance through 'Speaking Like an Indigenous State' and Respatializing Resources: Ethical Livelihood Concepts in Bolivia as Versatility or Verisimilitude?" *Geoforum 64*: 314–324.

# Part I

# Legal, socio-political and institutional contexts of extraction

# 1 Tendencies in tension

## Resource governance and social contradictions in contemporary Bolivia

*Tom Perreault*

Bolivia has undergone profound and far-reaching political, social, and economic transformation in recent years. A paragon of Washington Consensus-inspired neoliberalism in the 1980s and '90s, the country experienced the convulsive effects of widespread social protest and political upheaval in the early years of the twenty-first century, much of it centered on questions of resource governance (Perreault 2006). Beginning with the Cochabamba "water war" in 2000, an array of social movements led by the country's largest labor and peasant unions (the Bolivian Worker's Central (COB) and the Unified Syndical Union of Rural Workers of Bolivia (CSUTCB), respectively), together with irrigators' federations, urban neighborhood associations, and indigenous organizations, Bolivia's rural and urban poor masses rejected the neoliberalization of urban water services and natural gas, and demanded—not for the first time—that Bolivia's natural resources be used to benefit the Bolivian people, rather than foreign interests and national elites. Social movements overthrew the ruling oligarchy, forcing the resignation of neoliberal presidents in 2003 and again in 2005, clearing the way for the election in December 2005 of Evo Morales and his Movement to Socialism party.

Morales honed his political skill as the leader of the country's largest coca-growers' union (and retains this post today, as president of the Republic). He is also Bolivia's first president of indigenous descent—in a country with an indigenous majority. This story is by now well known (see, for example, Dunkerly 2007; Farthing and Kohl 2014; Gutierrez Aguilar 2014; Kaup 2013; Kohl and Farthing 2006; Postero 2007; Webber 2011). My concern here is with the legacy of these social and political transformations. In particular, I examine certain contradictory tendencies that have emerged in the governance of resource extraction (and in particular, in the governance of mining) in the decade since Morales's election. In doing so, I examine recent legislation that has bearing on mining, including the 2014 mining law (Ley de Minería y Metalurgia); the 2010 Law of the Rights of Mother Earth (Ley de Derechos de la Madre Tierra); the 2012 Law of Mother Earth and Integrated Development for Living Well

(Ley de Madre Tierra y Desarrollo Integral para Vivir Bien); and the 2013 Prior, Free and Informed Consultation Bill (Anteproyecto de Ley de Consulta Previa, Libre e Informada), which is currently being considered by Congress. My examination takes the form of a textual analysis of these laws, together with a political economic analysis of the context in which the laws are written and enacted. I undertake this examination fully aware of its limitations, and the fact that this approach tells us very little about the day-to-day lived experience of people in zones of extraction. Detailed ethnographic investigation of this sort is beyond the scope of the present chapter, but see Perreault (2013a, 2015, 2017, forthcoming), and Marston and Perreault 2017). The aim of the current chapter is to provide insight into the interplay of intellectual currents that inform legal frameworks.

The contradictions within and between these laws are apparent: the new mining law, arguably the most far-reaching and influential of the four, expands the mining sector, while favoring the interests of mining cooperatives and transnational firms, and weakening the ability of rural communities to safeguard their lands and waters. This law, written in large part by mining cooperatives and their allies, is widely seen as a boon to the mining sector and as weakening the rights of affected communities and others to contest mining activities. By contrast, the Law of Prior Consultation would establish the right of local communities to greater participation in decision-making regarding extractive activities, in accordance with international norms, thus potentially serving as a check on extractive interests. Of a rather different character is the Law of the Rights of Mother Earth, which draws on indigenous ideas of "living well" (*vivir bien* or *buen vivir* in Spanish, *sumak kawsay* in Quechua, *suma qamaña* in Aymara) to establish a normative philosophical basis for national development "in harmony and equilibrium with Mother Earth" (Art. 2). Observers within and beyond Bolivia have noted the obvious contradictions between the Bolivian government's ongoing intensification of resource extraction and its professed commitment to living harmoniously with Mother Earth. Read dialectically, however, it is apparent that the contradictory tendencies in these laws emerge from historically sedimented interests of particular social groups. These laws, then, may be read as legislative attempts to reconcile fundamental tensions within Bolivian society and political economy. In what follows, I examine these tensions and what they mean for the governance of mining in Bolivia. The next section examines three fundamental concepts at the heart of what Bolivians refer to as the "process of change" (*proceso de cambio*): neo-extractivism, post-neoliberalism, and *sumak kawsay/vivir bien*. This is followed by an overview of Bolivia's mining sector, and its political economic importance, both historically and in the current moment. The chapter then analyzes several pieces of legislation, recently passed or under consideration, and their relationship to the mining economy. The chapter ends by considering this legislation in the broader context of political economic and cultural political forces at play in the central Andean region.

## Antinomies of resource governance

During the past decade, the concept of environmental governance has emerged as a principal analytical lens through which social scientists have examined the political and economic coordination of socio-natural relations, particularly in the context of neoliberal capitalism (Bridge and Perreault 2009; Himley 2008; Lemos and Agrawal 2006). In a broad sense, environmental governance serves as an analytical frame to examine questions of rule and decision-making regarding nature and natural resources. In particular, critical studies in environmental governance focus on the shifting spatial scales and institutional arrangements of environmental management under neoliberalism. Within this perspective, authors have examined the governance of a range of resource sectors, including urban drinking water and sanitation (Bakker 2002, 2003; Jepson 2016; Mirosa and Harris 2012), mineral extraction (Bridge 2000; Himley 2013), forests (Prudham 2004), fisheries (Mansfield 2004; St. Martin 2005), peasant irrigation systems (Perreault 2005), and industrial agriculture (Hollander 2004). This work draws directly or indirectly on neo-Marxian approaches such as regulation theory to explain the organizational and institutional shifts associated with nature's commodification, enclosure, privatization, and marketization (Bakker 2010; Bridge and Jonas 2002). Commonly described using the shorthand govern*ment* to govern*ance*, this work describes the ways in which decision-making has been re-institutionalized, reorganized and rescaled under neoliberalism (for example through privatization of resource ownership or the establishment of public–private partnerships, market-based metrics for allocating resources, or various participatory mechanisms, each of which can operate across a variety of spatial scales) (Himley 2008). Of crucial importance here is the fact that, far from being rigid and durable socio-political configurations, the regulatory regimes that characterize modes of environmental governance arise out of social struggle and are thus contingent and unstable and must constantly be remade. In the mining sector especially, characterized as it is by dramatic boom-and-bust economic cycles and contentious labor relations, such regulatory regimes tend toward crisis and instability.

An environmental governance framework, most commonly associated with the analysis of nature's neoliberalization (Bakker 2010) may seem an uncomfortable fit with the putatively "post-neoliberal" states of Andean South America. In the case of the left-leaning Bolivian government of Evo Morales, a governance lens can illuminate the disjunctures as well as the continuities with the former neoliberal regime, through a focus on state/non-state relations, the re-institutionalization of resource management, and the shifting scales of rule and decision-making. In contemporary Bolivia, the social, economic, and political arrangements through which minerals (and hydrocarbons, water, lithium and forests, for example) are governed have been re-institutionalized and rescaled in the move away from neoliberalism, just as they were reconfigured in the transition *to* neoliberalism during the 1980s and 1990s. This post-neoliberal institutional reconfiguration has not involved merely a return to the pre-neoliberal *status quo ante*, however. Rather, the governance of resource extraction has been re-centralized through the so-called re-nationalization of the state hydrocarbons firm YPFB and the partial revival of the state mining company COMIBOL, even as political administration has been further *de*centralized through the creation of various forms of territorial autonomy (Farthing and Kohl 2014). Moreover, and of vital importance in the case of Bolivia, the state acts through numerous non-state actors. In his analysis of the post-neoliberal Bolivian state, for instance, Gustafson (2010) highlights the *Movimiento al Socialismo* (Movement to Socialism, MAS) government's efforts to re-signify state sovereignty in coalition with social movement actors and the use of social movement-like tactics, including mass mobilization and spectacle. In this sense, the Bolivian state extends its power in diffuse fashion through its relations with particular social blocs, among them coca growers, campesino unions, and mining cooperatives.

Of central importance to my argument is the basic fact that the institutional arrangements of environmental governance are seldom invented out of whole cloth. Rather, such arrangements are constructed, often in piecemeal fashion, in the context of pre-existing social, economic, and environmental topographies and must account for racial, class, and regional animosities, ongoing patronage relations, institutional dysfunction, popular resistance, and any number of other obstacles. As legal frameworks, governance arrangements are most often constructed in evolutionary fashion, that is, building on the institutional tools, capacities, and tendencies already in place. Moreover, as is true of nearly all legislation, environmental governance frameworks are the result of social and political struggle, and reflect either compromise positions or the class (and/or racial, gendered, regional or other) interests of dominant social groups. In this sense, legislation associated with resource governance may be read backward, in order to tease out the conflicts and compromises that gave rise to particular institutional arrangements. This is my intent in examining recent Bolivian legislation. First, however, I consider three concepts that are fundamental to understanding the current social and political moment in Bolivia: neo-extractivism, post-neoliberalism, and sumak kawsay.

## Tendencies in tension

Recent Bolivian legislation concerning environmental governance (either directly or indirectly), may productively be read through the conceptual lens of three political economic and cultural currents that shape public discourse and policy, and which reflect multiple, historically-rooted tensions, interests, and conflicts. These currents—neo-extractivism, post-neoliberalism and sumak kawsay ("living well")—are not opposed to one another (e.g., renewed resource extraction is not necessarily contrary to the principles of "living well"), but nor can they be reduced to elements of a single social or political body of thought. They are not in contradiction, in a Marxian sense, but rather exist in tension with one another, and as such may be brought into alignment only through the political and cultural labors of politicians, intellectuals and occasionally social movements. Here, I briefly discuss neo-extractivism, post-neoliberalism and sumak kawsay, as the economic, political and cultural frame for my subsequent discussion of recent legislation in Bolivia.

Since the 1990s, countries throughout Latin America, and particularly in the Andean region, have experienced a boom in commodities production and export. Driven in large part by sustained and spectacular economic growth in China (and, to a lesser extent, India and Brazil), Andean states have dramatically increased the production of petroleum, natural gas, and minerals, as well as certain agricultural commodities such as soy and palm oil, all destined for international markets. While the extractive "super-cycle" appears to be drawing to a close, with economic slowdown in China and full-blown crisis in Brazil, it has profoundly shaped Andean societies and political economies since the 1990s. Commodity production of this sort is not new to Latin America, of course. Spanish and Portuguese conquistadors came in search of gold and silver, and turned coastal areas (particularly in northeast Brazil) into vast plantations for the production of commercial crops such as sugar cane and cacao. Mining, in particular, has been historically central to the economies and politics of Peru, Bolivia and Chile. Since the outset of the twentieth century, however, the overall character of Andean mining has shifted in a variety of ways (Castillo 2013). During this period, mining has transformed from a labor-intensive to capital-intensive activity, with the shift to "mega-mining" favoring the dominance of transnational capital. In the 1980s and early '90s, roughly 12 percent of international mining investment flowed to Latin America, a figure that had increased to 33 percent by 2010 (Bebbington 2012; Bebbington and Bury 2013). Foreign participation in extractive industries in the Andean region was facilitated through the 1980s, '90s and early 2000s by a wave of neoliberal reforms designed to restructure property rights, taxes and royalty regimes (Himley 2010).

The recent intensification of resource extraction has been termed "neo-extractivism," mainly by its critics, as a way to signal both continuity with, and a slight shift away from, the long histories of Andean mining. Writing at the height of the commodities boom, Eduardo Gudynas (2009; see also Bebbington 2009, 2012), characterized neo-extractivism as (1) the renewed intensification of resource extraction in the context of favorable global commodity markets, combined with (2) efforts by states to capture a greater share of the rents generated by resource extraction and export, and (3) the redistribution of those rents through various forms of state-led social programs. In Bolivia, despite nationalist rhetoric to the contrary, neither the mining nor hydrocarbons sectors were truly nationalized. Rather, the state has assumed a larger role in administering exploration and extraction, and has renegotiated concession and rent structures so as to capture a greater share of revenue (to a notably greater degree in the hydrocarbons sector than in mining). These rents are redistributed through a variety of targeted cash transfer schemes (*bonos*), to aid school age children (*Bono Juancito Pinto*), mothers of small children (*Bono*

*Juana Azurduy*), and the elderly with minimal or no pensions (*Renta Dignidad*). As Gudynas (2009) notes, however, under "neo-extractivist" regimes, Bolivia, Peru, Ecuador, Colombia, and Chile have only deepened their long-standing reliance on the export of raw materials. With the drop in oil prices and the recent slowdown of the Chinese and Brazilian economies, an intensified "neo-dependency" has emerged as the other face of neo-extractivism (Díaz Cuellar 2015; Grugel and Riggirozzi 2012).

In "pink tide" states such as Bolivia and Ecuador, where voters have decisively rejected the neoliberal establishment and have elected leftist governments, neo-extractivism is closely related to the politics of post-neoliberalism (Peck, Theodore and Brennner 2010). If neo-extractivism is, in the main, an economic project, then post-neoliberalism has a decidedly political bent. Self-acknowledged post-neoliberal governments set themselves up in contrast not just to the neoliberal policies of their predecessors, but in opposition to (neo)imperialism and the legacies of European and North American domination. I would argue, then, that, like post-colonialism, we may understand post-neoliberalism as simultaneously encompassing an historical moment (the period *after* neoliberalism), a political project (closely allied to anti-colonial and anti-imperial movements), and a form of social subjectivity (which entails socio-cultural and political-economic identities). In this sense, post-neoliberalism is less a stable condition, or "thing-in-itself," and more a polyvalent and historically specific social and political process. As Grugel and Riggirozzi (2012, 3) explain it, post-neoliberalism entails a reconfiguration of the state vis-a-via capitalism and society, "based on a view that states have a moral responsibility to respect and deliver the inalienable (that is, not market-dependent) rights of their citizens, alongside growth." The authors go on to note that, politically, post-neoliberalism must be understood as a reaction against the excessive marketization of late capitalism, and the neglect of social need. Far from a rejection of (or attempt to de-link from) global capitalism, post-neoliberalism is thus an attempt to align the state's responsibilities for social inclusion—and in the cases of Bolivia and Ecuador, cultural recognition—with the exigencies of economic growth, in the context of a changing global political economy (Grugel and Riggirozzi 2012, 4). Given the path dependency of institutional conditions at both national and international scales, Yates and Bakker (2014) acknowledge that post-neoliberalism does not and indeed *cannot* entail a clean break with the neoliberal past. As such, post-neoliberal policies display both striking continuities with, as well as clear divergences from previous neoliberal frameworks, a structural condition that may in the end limit its own transformative potential (Kohl and Farthing 2012; Peck, Theodore and Brennner 2010). More reformist than revolutionary (pronouncements of Bolivian and Ecuadorian presidents notwithstanding), post-neoliberalism is characterized as (1) a utopian political project aimed at breaking with forms of foreign political, economic and cultural domination (including, but not limited to, neoliberalism); (2) various policies and practices designed to exert state control over market functions in order to address social concerns through redistributive measures; and (3) the deepening of democracy and social inclusion through various forms of regionally- and group-defined autonomy and self-government (Yates and Bakker 2014; see also Andreucci and Radhuber in press).

As Sarah Radcliffe has noted, while most analyses of post-neoliberalism skew toward the macroeconomic (see also Grugel and Riggirozzi 2012; Peck, Theodore, and Brenner 2010), in Andean Latin America the concept also raises important questions regarding the nature of citizenship, social belonging, development, and a "rights-based articulation of individual capacities and wellbeing, nature, and resource distribution" (Radcliffe 2012, 240). To the extent that post-neoliberalism is, at least in part, a politics of social inclusion and reinvigorated democracy, in Bolivia and Ecuador, with their strong indigenous movements, it aligns with the indigenous inspired idea of sumak kawsay. Glossed in Spanish as *vivir bien*, or *buen vivir* (in turn glossed in English as "living well"), the notions of sumak kawsay[1] (in Quechua) and *suma qamaña* (in Aymara) are concepts derived from Andean indigenous cosmology. Zimmerer (2012) calls *kawsay* a "portmanteau" concept and one of the most versatile words in the Quechua lexicon, whose meanings in colonial Peru ranged from basic notions of existence to normative evaluations of wellbeing and health. By the twentieth century, kawsay was associated with traditional indigenous foodways and agricultural practices, and more recently with agroecology, food security, sustainable development, and social justice (ibid). It would be a mistake, however, to assume that the concept has been adopted in unmediated form by Andean lawmakers, as politicians in Bolivia and (especially) Ecuador have selectively appropriated the language of sumak kawsay, engaging with and at times coopting elements they find politically expedient, while ignoring (or showing open hostility toward) others (de la Cadena 2010). The concepts of sumak kawsay, suma qamaña and vivir bien/buen vivir are closely related to the Quechua and Aymara concept of *pachamama*, glossed as *madre tierra* in Spanish and earth mother/Mother Earth in English. I would argue that the concept of pachamama is a more expansive concept than the English notion of "nature," and encompasses humans in relation to the communities—both human and non-human–in which they live (Zimmerer 2015).

Rather than assessing how closely the Bolivian government's uses of sumak kawsay resemble the concept's supposedly pristine indigenous meanings—surely a fool's errand—it is more analytically productive to consider the political work the concept does in contemporary Bolivia, and particularly in relation to neo-extractivist policies. As Zimmerer (2015) notes, the use of indigenous concepts such as sumak kawsay provides political legitimacy for Evo Morales and his MAS party. For instance, Article 8 of the 2009 Constitution recognizes Bolivia as a plural society, founded on the principle of suma qamaña, and the Andean ideals of *ama quilla, ama llulla, ama suwa* (do not be lazy, do not lie, do not steal). The adoption of indigenous concepts into the Constitutions and various statutory legal frameworks of Bolivia and Ecuador reflect the long-term struggles of indigenous social movements, dating to the mid-twentieth century, with the formation of indigenous/campesino organizations (frequently associated with agrarian reform and associated struggles for land rights and territorial recognition). By the 1980s and '90s, indigenous mobilization in both countries involved periodic mass protest, which forced the resignation of multiple neo-liberal presidents in each country. As well, indigenous people maintained a public and political presence through the formation of strong national-level social movement organizations (most notably CONAIE in Ecuador and CSUTCB, CONAMAQ and CIDOB in Bolivia[2]), as well as political parties that allowed participation in electoral politics. In Bolivia, this mobilization culminated with the election of Evo Morales in December 2005. The codification of indigenous concepts such as sumak kawsay and pachamama reflects an effort at cultural inclusion and democratic deepening to account for the values of the historically marginalized population that constitutes the majority in Bolivia and a substantial minority in Ecuador. While there is little doubt that the concepts are used selectively and are highly mediated through a contemporary lens (de la Cadena 2010), it is also true that the concept of sumak kawsay resides at the "cultural borderlands" between indigeneity and dominant capitalist society, which imbues it with an elasticity of meaning that translates across cultural and linguistic divides (Zimmerer 2012). The concept's resonance with mainstream (and similarly flexible) notions of sustainable development, food sovereignty, and multi-culturalism facilitate its broad adoption and malleable interpretation.

To summarize, contemporary resource governance in Bolivia is best viewed within the context of three over-arching and historically constituted political economic and cultural currents: neo-extractivism, post-neoliberalism and vivir bien/sumak kawsay. These concepts establish a conceptual frame for under-standing the institutional arrangements governing mining and hydrocarbons development. Importantly, they are neither necessarily in contradiction, nor are they reducible to one another. Rather, they coexist in tension, and their alignment requires the ideological labor of politicians, intellectuals and social movements. The core question for this chapter, then, concerns the political and cultural work these concepts are made to do in the service of resource governance. In what follows, I examine several pieces of Bolivian legislation that have bearing on resource governance, drawing on the concepts of neo-extractivism, post-neoliberalism and sumak kawsay as a conceptual framework.

## Legal frameworks for governing extraction

Today we export three times more in volume than the year 2005 and this speaks of a mining country. Bolivia lives off its gas, but also off mining and we are proud of it.

(Vice President Álvaro García Linera 2012, 32; cited in Díaz Cuellar 2015)

Since taking office in 2006, Evo Morales and his MAS party have set out to transform resource governance in Bolivia. This has entailed the passage of numerous laws and decrees intended to codify the MAS's vision of a political and economic system rooted in Andean indigenous values—a variation of "21st Century Socialism" shared with Ecuador and Venezuela—and based on the extraction and (mostly) export of raw materials. Among these, by far the most important is natural gas, produced in the dry, lowland Chaco region in the country's southeast. Bolivia produces oil as well, though to date its production has been consumed internally rather than exported. This situation may change if current efforts to exploit petroleum reserves in northern La Paz department— in the heart of the country's Amazon rainforest (and including in protected areas and within indigenous territories)—prove successful. Bolivia has roughly half the world's proven reserves of lithium, in the Uyuni salt flats of the southern Altiplano, but efforts to exploit them have been slow and have yet to provide revenue to the state.

The Morales government has sought selectively to extend state influence in the important sectors of gas and mining. In May 2006, shortly after taking office, one of Morales's first major initiatives was the "nationalization" of natural gas operations, under the "Heroes del Chaco" decree. The title of the decree invokes those who died in the ill-fated Chaco War against Paraguay (1932–1935), in which Bolivia lost much of its Chaco territory, where most of its natural gas is now produced. Thus, Morales's move to nationalize gas operations is symbolically tied, historically and geographically, to a sense of heroism, sacrifice and nationalist sovereignty (Perreault 2013b; Perreault and Valdivia 2012). It is worth noting, however, that the Heroes del Chaco decree was in fact less a true nationalization than it was a forced renegotiation of contracts with the transnational hydrocarbons firms on which Bolivia relies to extract and export the gas. The decree restored royalties to their pre-neoliberal rates and reconstituted the state hydrocarbons firm YPFB (which had been removed from upstream activities during the 1980s and '90s), and reestablished its role as a partner in exploration and extraction (Kaup 2013).

Similar, though less ambitious, arrangements were established a few months later in the mining sector. In October 2006 a conflict broke out at the Huanuni mine in Oruro, between miners working with mining cooperatives and salaried miners working for the state mining firm COMIBOL, both of whom claimed access to the mine's richest tin deposits. Conflict that began belowground emerged at the surface and led to 17 deaths and over 100 injuries (Howard and Dangl 2006; López, et al. 2010). Coming just a few months after the "nationalization" of the gas sector, the state's response was to nego- tiate the absorption of mining cooperatives into COMIBOL, as part of the latter's Huanuni Mining Company. This led to a more than six-fold increase in Huanuni's workforce (from about 700 to over 4500), both facilitating and necessitating a dramatic increase in production (Perreault 2013a). The expansion and reassertion of COMIBOL's control over the Huanuni mine paved the way for similar (though less ambitious) moves in other mining operations, including the Vinto smelter (Oruro) in 2007, and the Colquiri mine (La Paz), brought under state control following violent conflict in 2012.

The resuscitation of YPFB and COMIBOL, and the state's reassertion of control in the hydrocarbons and mining sectors are emblematic not just of Bolivia's post-neoliberal turn in resource governance, but also of the intensification of Bolivia's dependence on resource extraction. According to the Bolivian National Institute of Statistics, hydrocarbons and mining have together increased as a share of Bolivia's exports, from less than 50 percent of exports by value (at roughly US$1.5 billion) in 2000, to over 80 percent (and roughly US$12 billion) in 2014. If agricultural exports (mostly soy and beef) are considered, primary commodities currently account for 96 percent of Bolivian exports (Díaz Cuellar 2015). The renewed emphasis on resource extraction has been both motivated and facilitated by sustained demand from China and other emerging economies, as well as by nationalist politics aimed at overturning the ruinous neoliberal policies of previous decades. Moreover, the legal codification of resource governance reflects the three political currents discussed above: neo-extractivism, post-neoliberalism, and sumak kawsay. In what follows, I examine recent laws, in order to trace the political currents they represent.

### Neo-extractivism meets post-neoliberalism: the 2014 Law of Mining and Metallurgy

Bolivian analysts generally agree that the 2014 Law of Mining and Metallurgy (Law No. 535) largely reflects and reinforces the interests of the country's mining cooperatives. Representatives of the national federation of mining cooperatives (FENCOMIN) were closely involved in its development, along with representatives of private mining firms and COMIBOL. These three groups are formally recognized in Article 369 of the 2009 Constitution as the mining sector's "productive actors"—a list that notably excludes community mining, which at times has been the target of criticism and protest by the highly organized mining cooperatives. FENCOMIN sought legal approval for potentially lucrative contracts with private firms, of the sort that some of its wealthiest and most influential member cooperatives had enjoyed for some time. Indeed, in the original draft of the bill, this was precisely the intent of Article 151. In April 2014, when Congress changed the wording of the article to expressly prohibit such arrangements as unconstitutional, FENCOMIN broke with the government and called on its members to set up road blockades throughout the country in protest (Francescone 2015). The cooperatives lost this particular battle; congressional revisions to Article 151 remained in the final text, prohibiting cooperatives from forming contracts directly with private mining firms.[3] Nevertheless, the law formalized and in some cases substantially expanded many of the privileges cooperatives had long enjoyed, including their classification as not-for-profit social organizations (Art.127), which confers on them nearly tax-free status. The law also grants cooperatives the right to maintain existing contracts with private firms and subcontract their concession areas to third parties (Art. 130), and form mixed companies ("empresas mixtas") with public companies via COMIBOL (Art. 151).

In spite of the prohibition on forming new contracts between private firms and mining cooperatives, these two groups are widely seen as the biggest beneficiaries of the new mining law. Indeed, many of the government's critics argue that the law does too little to reconstitute COMIBOL and too much to cater to the interests of cooperatives and private capital. Early drafts of the law were also criticized for failing to ensure environmental protections and community rights (Campanini 2012). In the context of the conceptual framework outlined above, the 2014 mining law may be read as reflecting the state's commitment to neo-extractivism and post-neoliberalism. While intended primarily to stimulate the mining sector, and particularly mining cooperatives and private mining firms—viewed by the government as the most dynamic actors in the sector—the mining law also goes some way toward strengthening state control over mineral resources and mining governance. A clear example of this is the promotion of joint ventures between cooperatives and the state (Art. 151). The reassertion of state control is also reflected in Article 2, which establishes mineral resources as the property of the Bolivian people, and their administration as the responsibility of the Bolivian state. Similarly, Article 39 establishes a semi-autonomous legal entity for governing the mining sector (the Autoridad Jurisdiccional Administrativa Minera, AJAM). Operating under the authority of the Ministry of Mines and Metallurgy, AJAM is responsible for administering mining contracts and adjudicating disputes over mining claims. Ironically, the legal and political structure of the AJAM is similar to the semi-autonomous "superintendencies" (*superintendencias*) established during the neoliberal presidency of Gonzalo Sánchez de Lozada, and charged with administering various utilities and resource sectors. During the late 1990s and early 2000s, these unelected, semi-autonomous bureaucracies were widely seen as serving the interest of transnational capital and became targets of anti-neoliberal protest (Perreault 2005). Morales dismantled the superintendencies shortly after taking office. That a *post*-neoliberal resource law establishes a nearly identical administrative structure for the governance of minerals and mining speaks to the continuities between neoliberalism and the various forms that post-neoliberalism takes (Peck, Theodore, and Brenner 2010; Yates and Bakker 2014).

While the mining law does reference the Andean indigenous value of reciprocity with mother earth ("*Reciprocidad con la Madre Tierra*"—Art. 5, on principles of the law) and the principles of vivir bien (Art. 208, on prior consultation), it is notably vague about the specific responsibilities of mining operators with regard to protecting indigenous rights and environmental quality. In a law that runs to 95 pages with 234 articles, just two articles (Arts. 217 and 218), together covering less than half a page, are dedicated to environmental responsibilities. By contrast, 16 articles are dedicated to the structure and functions of the administrative agency AJAM, another 17 detail the various mining companies in operation, and 22 articles are devoted to the rights of mine operators. The environmental articles simply state that mining operators must abide by existing environmental law, and that operators must obtain an environmental license from the competent environmental authority. Article 222 establishes that the Ministry of Mining and Metallurgy (notably *not* the Ministry of Environment) will oversee mining operations and ensure compliance with environmental laws. The most widespread and acute forms of environmental contamination associated with mining operations affect surface waters. Acid mine drainage, releases of contaminated water from processing plants, accidental spills, and intentional dumping affect rivers and lakes, extending the footprint of mining operations for miles downstream (Bridge 2004). The mining law contains just three Articles regarding water (Arts. 111, 112, and 113), which are focused on the rights of miners to surface and sub-surface waters for use in mining operations. Without giving specifics, the law seems intentionally vague: Article 111 states that mining operations should not cause environmental damage and should comply with existing environmental laws.

Article 208 of the law establishes the right of affected communities to free, prior and informed consultation, in accordance with the Constitution and the pending Prior Consultation bill. The right to consultation in the mining sector is highly circumscribed, however, and is limited to "indigenous-campesino nations and peoples, intercultural communities and Afro-Bolivian peoples" (see Perreault 2015). While ensuring the right of these vulnerable groups to consultation is surely laudable, this designation excludes the majority of rural communities and urban neighborhoods actually affected by mining activity. Article 209 establishes that only those communities of pre-colonial origin, living in ancestral territories and maintaining traditional ways of life have a right to prior consultation. Given the disruption of the Colonial and Republican eras, and widespread processes of rural to urban migration, these criteria pertain to relatively few communities—even in a country with an indigenous majority. The right to consultation is further limited by Article 207, paragraph III, which states that prior consultation is required only in the case of new mining contracts, rather than new activities within existing contracts, thus excluding most mining operations in the major mining districts of Potosí and Oruro. As I outline below, the aims of the 2014 mining law stand in sharp contrast with those of the Law of Mother Earth and the Law of the Rights of Mother Earth, examples of legislation promoting the values of sumak kawsay and vivir bien.

### Legislating Living Well: sumak kawsay and the Mother Earth laws

In December 2010, early in his second term as president, Morales signed into law the Ley de Derechos de la Madre Tierra (Law of the Rights of Mother Earth). The text of the law is short—just five pages—and general in its orientation. It provides a generalized and in places elegant summary of beliefs that can be glossed as "indigenous cosmovision"—a term of cultural translation frequently used to describe beliefs and practices reflective of a generalized indigenous culture. In this sense, the language of indigenous cosmovision provides a "middle ground" that serves both to unify indigenous subjectivities and articulate them with national political positions (Conklin and Graham 1995; see also Albro 2005). The challenge for indigenous activists and politicians in Bolivia and elsewhere is to make indigenous political subjectivities legible to non-indigenous individuals within liberal legal–political institutional frameworks (Van Cott 2007). More political statement than legal instrument, the Law of the Rights of Mother Earth should not be read as "environmentalist" in any straightforward sense. It is, rather a conceptual framework for cultural translation, which articulates ideas prevalent in cosmopolitan environmentalism with an Andean understanding of indigeneity, conceived of as rooted in reciprocity, complementarity and harmony in socio-natural relations. Article 1 states the Law's intention, which is to "recognize the rights of Mother Earth as well as the obligations of the Plurinational State [of Bolivia] and society to guarantee respect for those rights."

Article 2 details a set of principles in which the law is based, including (1) harmony; (2) collective wellbeing (*bien colectivo*); (3) a guarantee of the regeneration of Mother Earth; (4) respect and defense of the rights of Mother Earth; (5) rejection of privatization and commodification of natural resources (*no mercantilización*); and (6) interculturality (*interculturalidad*). Thus, the law seeks to articulate an Andean understanding of indigeneity ("the framework of plurality and diversity," Art. 2.1), and a rejection of neoliberal capitalism ("cannot be commodified (*mercantilizados*) . . . nor form part of the anyone's private inheritance (*patrimonio privado*)," Art. 2.5) with a sense of cosmopolitan environmentalism ("the State . . . should guarantee the necessary conditions for the regeneration of the diverse systems of Mother Earth . . . understanding that those systems are limited in their capacity to regenerate," Art. 2.3). The relationship between Mother Earth and indigenous cultures is further developed in subsequent articles. In this way, the Law places the notion of Living Well at the center of an idealized Bolivian way of life.

Two years after the passage of the Law of the Rights of Mother Earth, the ideas it contained were expanded and further developed in the Law of Mother Earth and Integrated Development for Living Well (Ley de Madre Tierra y Desarrollo Integral para Vivir Bien, passed into law on 15 October 2012). At 43 pages, this is a more fully developed and detailed statement on socio-natural relations, development ideals and the concept of living well. Article 1 clearly lays out the Law's purpose: "The objective of this Law is to establish the vision and foundations of integrated development (*desarrollo integral*) in harmony and equilibrium with Mother Earth for Living Well (*Vivir Bien*), guaranteeing the continuity of the regenerative capacity of Mother Earth's components and life systems, recovering and strengthening the local and ancestral knowledges, in the context of complementary rights, obligations and duties; as well as the objectives of integral development as a means to achieve Living Well . . ." Much of the Law restates elements of the earlier Law of the Rights of Mother Earth, including the establishment of the "conditions necessary for a transition to Living Well in harmony and equilibrium with Mother Earth"

(Art. 3.3): rejection of the commodification of the environmental functions of Mother Earth (Art. 4.2), and the promotion by the state of "harmonious, adaptive and balanced relations between the needs of the Bolivian people and the regenerative capacity, components and life systems of Mother Earth" (Art. 4.12).

In comparison with the 2010 Law of the Rights of Mother Earth, the 2012 law places greater emphasis on cultural and economic diversity. Article 4.4 even specifies that small-scale miners and mining cooperatives—notorious for poor environmental records—will work with state authorities to prevent damage to Mother Earth. Article 4.15 draws on a notion of Andean indigeneity in representing a "plural economy" (*economía plural*):

> The Plurinational State of Bolivia recognizes the plural economy as the Bolivian economic model, considering the different forms of economic organization, based on the principles of complementarity, reciprocity, solidarity, redistribution, equality, sustainability, balance and harmony, in which the social, community-based economy complements the individual interests with collective Living Well.

Similarly, Article 4.17 ("Diálogo de Saberes") states that the Bolivian state "recognizes the complementarity between traditional knowledges (*saberes y conocimientos tradicionales*) and science."

The recognition and promotion of cultural diversity is perhaps stated most plainly and poetically in Article 5.2 ("El Vivir Bien"), a term which is then translated into Aymara, Quechua and Guaraní, respectively, as Sumaj Kamaña, Sumaj Kausay, and Yaiko Kävi Pave. According to this article, Living Well is,

> the civilizational horizon and cultural alternative to capitalism and modernity that emerges from the cosmovisions of the indigenous original peasant peoples and nations, and the intercultural and afrobolivian communities, and is conceived in the context of interculturality . . . It means living in complementarity, in harmony and balance with Mother Earth and with societies in equity and solidarity, and eliminating the inequalities and mechanisms of domination. It is Living Well between ourselves, Living Well with that which surrounds us, and Living Well with oneself.

Article 5.3 goes on to note that integrated development for Living Well "is not an end in itself, but rather an intermediary phase, to achieve Living Well as a new civilizational and cultural horizon." Article 9 establishes rights under the Law, with Art. 9.2 recognizing the collective and individual rights of indigenous original peasant peoples and nations and intercultural and afrobolivian communities according to the Constitution, the United Nations Declaration on the Rights of Indigenous Peoples, and Convention 169 of the International Labor Organization, regarding the rights of indigenous peoples. Article 13.10 calls for the:

revalorization and strengthening of systems of life of smallholder farmers, of indigenous original peasant nations and peoples, intercultural and afrobolivian communities, cooperatives and other associative systems, according to sustainable management of their biodiversity and to respect, revalorization and reaffirmation of their knowledges in the context of cultural diversity.

Such examples abound in the Law's 58 articles. Clearly, the law highlights Bolivia's cultural diversity and promotes an Andean-centered vision of indigeneity centered on complementarity, reciprocity and balanced, harmonious social and socio-natural relations. It is worth drawing special attention to Articles 26 and 27, which address extractive industries (mining and hydrocarbons) and water, respectively. Here, the Law resembles an industrial guide for environmental best practices, highlighting the need for up-to-date and clean technology in exploration, extraction, processing and transport, in order to minimize environmental and social damage (Art. 26.1). Subsequent paragraphs in Article 26 concern the participation of indigenous and other communities (Art. 26.2); require the mitigation of environmental damage (Art. 26.4); and call for penalties for irreversible environmental damage (Art. 26.5). Article 27 establishes the need for safeguarding water in order to satisfy the needs of both human domestic uses and the conservation of ecological systems. Importantly, paragraph 1 of the Article notes that these should be maintained in order to guarantee food security and sovereignty (*seguridad y soberanía alimentaria*). The Article calls for extractive industries and industrial plants to implement measures in order to minimize their contamination of water resources (Art. 27.2), and promotes watershed-based management of water resources (Art. 27.8). In this way, the law serves to integrate indigenous cosmovision with the language of cosmopolitan environmentalism.

A third piece of legislation worth considering here is the Free, Prior and Informed Consultation Bill (Anteproyecto de Ley de Consulta Previa, Libre e Informada). Drafted in 2013, the Bill is still under consideration in parliament, though many of its articles have been approved by the national legislative commission that considered the Bill prior to parliamentary debates. The Bill reflects international efforts by UN agencies, multilateral lenders, development agencies, human rights advocates, and others to implement standardized policies of free, prior and informed consent (known by the acronym FPIC). The concept of prior consent (*consentiamiento previa*) is contained in Article 3 of the Bill, but it is the weaker standard of prior consultation (*consulta previa*), for which the Bill is named, which is highlighted in the Bill as well as in existing statutory laws regulating hydrocarbons and mining (Perreault 2015). The Bill has widespread support in Bolivia, both for the implications it holds for cultural and human rights and for its international appeal as a component of modern resource extraction. Because it is likely to undergo further revision by parliament before it becomes law, my analysis here will be brief. It is worth noting, however, that the law focuses specifically on the collective rights of indigenous people, and as such it articulates with understandings of vivir bien. Article 2 of the Bill states,

> The right to prior, free and informed consultation is a collective and fundamental right of the indigenous original peasant nations and peoples, intercultural communities and afrobolivian peoples . . . It is a democratic mechanism for the exercise of collective rights and in particular free determination as well as to deepen direct, participatory and community democracy.

Article 3 of the Bill defines prior consultation as a means to assure that the development aspirations of indigenous and peasant peoples are incorporated into planning efforts. This should be achieved through intercultural dialogue between the state and indigenous original peasant nations and peoples, and intercultural and afrobolivian communities, prior to final decisions regarding extractive projects and in order to harmonize different visions of development and "in order to achieve Living Well" (*para alcanzar el Vivir Bien*). The Bill goes on to call for respect for Mother Earth (Art. 4f), and specifically references ILO 169 as the international legal framework that lends FPIC its authority (Art. 37).

The point here is that the Free, Prior and Informed Consultation Bill, like the Mother Earth laws, articulates with the Bolivian state's vision of Living Well, which itself is derived from the Andean indigenous concepts of sumak kawsay/suma qamaña. These concepts reflect an explicit effort by Evo Morales and the MAS government to recognize Bolivia as a plurinational state, and to valorize indigenous cultural forms, linguistic diversity, belief systems, and plural forms of socioeconomic organization. These efforts emerged through decades of struggle by Bolivia's indigenous majority, which until Morales was elected president was largely excluded from political power. The recurring tropes of "Living Well" and "Mother Earth," and the frequent references to "indigenous original peasant nations and peoples, and intercultural communities and afrobolivian peoples" (*naciones y pueblos indígena originario campesino y comunidades interculturales y pueblo afroboliviano*) serve to normalize these concepts—both in the sense of making them a "normal" part of legal discourse and in the sense that their promotion represents a normative reversal of historically sedimented power relations within Bolivia. In this way, these concepts are discursively linked to calls for cultural and political sovereignty, and a rejection of colonialism, both external and internal.

## Conclusion

As I hope to have shown through my analysis of the 2014 mining law, on the one hand, and the two Mother Earth laws and the prior consultation Bill on the other hand, there are legal tensions at play within Bolivian resource governance. My argument is not that these tensions represent a form of hypocrisy per se (although statements by both Evo Morales and Vice President Álvaro García Linera provide plenty of evidence for such a claim). Nor am I arguing that these laws, and the social and political currents they represent, are fundamentally in contradiction. There is no inherent contradiction between mining and indigenous rights. My point, rather, is that the intensification of mining and the promotion of the interests of mining cooperatives and transnational mining firms exist in uneasy tension with efforts to expand the rights of indigenous, peasant and afrobolivian peoples who are often excluded from the benefits of resource extraction.

These legislative efforts may be framed conceptually by the ideas of neo-extractivism, post-neoliberalism and sumak kawsay/vivir bien. These concepts coexist in dialectical tension, informing one another and providing a conceptual framework for understanding political, economic and cultural processes underway in contemporary Bolivia. Neo-extractivism, as discussed here, is the renewed intensification of resource extraction, coupled with social policies aimed at transferring resource rents toward the poor and marginalized. As critics such as Gudynas (2009), Acosta (2008) and Bebbington (2009, 2012) have pointed out, the putatively post-neoliberal governments of Bolivia and Ecuador have intensified their commitment to resource extraction and by extension have deepened their dependence on global capitalism and, in particular, international demand for primary commodities. Whereas neo-extractivism is an essentially economic concept, the notion of post-neoliberalism is rooted firmly in the political. As such, it may be seen as a utopian project of governance, which rejects foreign political and economic domination and seeks a greater role for the state in economic activities, while placing greater emphasis on social inclusion and democratization (Yates and Bakker 2014). Of a somewhat different valence is the concept of sumak kawsay or "living well," a flexible "portmanteau" concept derived from Andean indigenous culture, which has been used to signify the rights of people in relation to their human and non-human communities (Zimmerer 2015). Sumak kawsay has been deployed selectively by Andean governments as a way to discursively acknowledge their commitment—however fragile—to interculturality and plurinationalism.

The common thread running through these concepts is that of social inclusion and economic development, broadly conceived. A potential (though not inherent) contradiction in these concepts is between neo-extractivism and the notion of sumak kawsay, with its connections to pachamama and the rights of non-human nature. Indeed, principles of sumak kawsay have been invoked in critiques of the extractivist policies of Andean states (e.g., Acosta 2008; Gudynas 2009). Indigenous peoples and the principle of sumak kawsay are not necessarily opposed to resource extraction—as de la Cadena (2010) has pointed out, mining has been part of Andean indigenous culture for centuries—but the massive environmental impacts associated with contemporary mining are difficult, if not impossible, to reconcile with any definition of "living well" or the rights of Mother Earth (see also Bridge 2004; Perreault 2013a).

Moreover, it must be acknowledged, as McNeish (2015) suggests, that the concepts of sumak kawsay, post-neoliberalism and neo-extractivism are projects of intellectual and political elites, and in the abstract have little resonance with the large majority of Bolivia's indigenous population. We may reasonably ask, then, what political work these concepts are made to do in the service of resource governance. However problematically, the concepts serve to articulate state projects with forces of global capitalism, on the one hand, and the historical demands of the country's indigenous campesino majority, on the other. Post-neoliberalism and neo-extractivism describe the state's strategy of gaining greater control over resource governance and capturing a greater share of resource rents, while distributing those rents more equitably among the population. They are, in essence, political economic projects. Sumak kawsay, by contrast, articulates a set of cultural concerns both to international audiences in solidarity with the Left governments of South America and, more importantly, with Bolivia's indigenous masses. The concept signals that the government has broken with the colonial, neoliberal past and has adopted the language (if not the full political program) of indigenous intellectuals and their organizations. The irony of this position is particularly evident in light of the MAS government's recent tensions with the indigenous organizations CONAMAQ (Consejo Nacional de Ayllus y Markas de Qullasuyu, National Council of Ayllus and Markas of Qullasuyu) and CIDOB (Confederación de Pueblos Indígenas de Bolivia, Confederation of Indigenous Peoples of Bolivia), in which the government effectively divided the organizations and put in power leaders sympathetic to the MAS (McNeish 2015). These tensions point to broader political, economic and cultural forces at work in the Andes that reinforce Bolivia's historical dependency on resource extraction (Kaup 2010; Kohl and Farthing 2012). Given that the global boom in commodities demand appears to have turned bust, this path dependency, which has only intensified in recent years, raises crucial questions regarding the continued viability of the neo-extractivist model. With oil prices at historic lows and demand for basic commodities waning in China, Brazil and elsewhere, the future of Andean economies looks uncertain. There is little doubt that the region's dependence on resource extraction will continue into the foreseeable future, but the conceptual armature of neo-extractivism, post-neoliberalism and sumak kawsay, as an ideological framework for resource governance, is increasingly tenuous.

## Notes

1    This is a Quechua term and is spelled in multiple ways, including sumak kawsay, sumaq kawsay, and sumaj kausay. To avoid confusion, I have adopted the spelling that most commonly appears in the academic literature.
2    CONAIE is the Confederación de Nacionalidades Indígenas del Ecuador (Confederation of Indigenous Nationalities of Ecuador), the main national-level indigenous organization in the country; CSUTCB is the Confederación Sindical Única de Trabajadores Campesinos de Bolivia (the Unified Syndical Confederación of Peasant Workers of Bolivia), the country's largest campesino union, with influence primarily in the Andean

west; CONAMAQ is the Consejo Nacional de Ayllus y Markas de Qullasuyu (National Council of Ayllus and Markas of Qullasuyu), an organization of indigenous/*originario* people in Bolivia's Andean region; CIDOB is the Confederación de Pueblos Indígenas de Bolivia (Confederation of Indigenous Peoples of Bolivia, formerly the Confederación de Pueblos Indígenas del Oriente Boliviano, Confederation of Indigenous Peoples of the Bolivian East), an organization of indigenous peoples in Bolivia's eastern and northern lowlands.

3    Cooperativistas renewed their protests in August 2016, blocking linking Oruro and La Paz departments. Five people died in the violence, including the vice-minister of the interior Rodolfo Illanes, who was beaten to death by miners when he attempted to negotiate an end to the blockades.

# References

Acosta, Alberto. 2008. "El 'buen vivir' para la Construcción de Alternativas". *Casa de las Américas 251* (April–June): 3–9.

Albro, Robert. 2005. "The Indigenous in the Plural in Bolivian Oppositional Politics." *Bulletin of Latin American Research 24* (4): 433–453.

Andreucci, Diego, and Isabella M. Radhuber. In press. "Limits to 'Counter-Neoliberal' Reform: Mining Expansion and the Marginalization of Post-Extractivist Forces in Evo Morales' Bolivia." *Geoforum.*

Bakker, Karen. 2002. "From State to Market? Water *Mercantilización* in Spain." *Environment and Planning A 34* (5): 767–790.

Bakker, Karen. 2003. *An Uncooperative Commodity: Privatizing Water in England and Wales.* Oxford: Oxford University Press.

Bakker, Karen. 2010. *Privatizing Water: Governance Failure and the World's Urban Water Crisis.* Ithaca: Cornell University Press.

Bebbington, Anthony J. 2009. "The New Extraction: Rewriting the Political Ecology of the Andes?" *NACLA Report on the Americas 42* (5): 12–20.

Bebbington, Anthony J. 2012. "Underground Political Ecologies: The Second Annual Lecture of the Cultural and Political Ecology Specialty Group of the Association of American Geographers." *Geoforum 43* (6): 1152–1162.

Bebbington, Anthony J., and Jeffrey Bury, eds. 2013. *Subterranean Struggles: New Dynamics of Mining, Oil, and Gas in Latin America.* Austin: University of Texas Press.

Bridge, Gavin. 2000. "The Social Regulation of Resource Access and Environmental Impact: Production, Nature and Contradiction in the U.S. Copper Industry." *Geoforum 31* (2): 237–256.

Bridge, Gavin. 2004. "Contested Terrain: Mining and the Environment." *Annual Review of Environmental Resources 21* (29): 205–259.

Bridge, Gavin, and Andrew E.G. Jonas. 2002. "Governing Nature: The Reregulation of Resource Access, Production, and Consumption." *Environment and Planning A 34* (5): 759–766.

Bridge, Gavin, and Tom Perreault. 2009. "Environmental Governance." In *Companion to Environmental Geography*, edited by Noel Castree, David Demeritt, Diana Liverman and Bruce Rhoads, 475–497. Oxford: Blackwell.

Campanini, Oscar. 2012. "El Agua: Entre los Conflictos Mineros y las Propuestas Normativas." *PetroPress 29* (July–September): 10–17.

Castillo, Gerardo. 2013. "Spatial Production of the Andes and Mining Historical Developmentin Peru." Societas Consultora de Análisis Social. *Acerca de Societas.* Web journal. February 17. https://societasconsultora.wordpress.com/2013/02/17/spatial-production-of-the-andes-and-mining-historical-development-in-peru/.

Conklin, Beth A., and Laura R. Graham. 1995. "The Shifting Middle Ground: Amazonian Indians and Eco-Politics." *American Anthropologist 97* (4): 695–710.

de la Cadena, Marisol. 2010. "Indigenous Cosmopolitics in the Andes: Conceptual Reflections beyond 'Politics.'" *Cultural Anthropology 25* (2): 334–370.

Díaz Cuellar, Vladimir. 2015. "Mining under the MAS Government: Transnational Companies, Cooperatives and the Commodity Boom in Bolivia." Paper presented at the Latin American Studies Association International Conference, San Juan, Puerto Rico, May 27–30.

Dunkerley, James. D. 2007. *Bolivia: Revolution and the Power of History in the Present.* London: Institute for the Study of the Americas.

Farthing, Linda, and Benjamin Kohl. 2014. *Evo's Bolivia: Continuity and Change.* Austin: University of Texas Press.

Francescone, Kirsten. 2015. "Cooperative Miners and the Politics of Abandonment in Bolivia." *The Extractive Industry and Society 2* (4): 746–755. http://dx.doi.org/10.1016/j.exis.2015.10.004.

García Linera, Álvaro. 2012. *Las Empresas del Estado: Parimonio Colectivo del Pueblo Boliviano.* La Paz: Vicepresidencia del Estado Plurinacional de Bolivia.

Grugel, Jean, and Pía Riggirozzi. 2012. "Post-Neoliberalism in Latin America: Rebuilding and Reclaiming the State after Crisis." *Development and Change 43* (1): 1–21.

Gudynas, Eduardo. 2009. "Diez Tesis Urgentes sobre el Nuevo Extractivismo: Contextos y Demandas bajo el Progresismo Sudamericano Actual." In *Extractivismo, Política y Sociedad*, 187–225. Quito: Centro Andino de Acción Popular (CAAP) and Centro Latino Americano de Ecología Social (CLAES).

Gustafson, Bret. 2010. "When States Act Like Movements: Dismantling Local Power and Seating Sovereignty in Post-Neoliberal Bolivia." *Latin American Perspectives 37* (4): 48–66.

Gutierrez Aguilar, Raquel. 2014. *Rhythms of the Pachakuti: Indigenous Uprising and State Power in Bolivia*, translated by Stacy Alba S. Skar. Durham, NC: Duke University Press.

Himley, Matthew. 2008. "Environmental Governance: The Nexus of Nature and Neoliberalism." *Geography Compass 2* (2): 443–451.

Himley, Matthew. 2010. "Global Mining and the Uneasy Neoliberalization of Sustainable Development." *Sustainability 2* (10): 3270–3290.

Himley, Matthew. 2013. "Regularizing Extraction in Andean Peru: Mining and Social Mobilization in an Age of Corporate Social Responsibility." *Antipode 45* (2): 394–416.

Hollander, Gail M. 2004. "Agricultural Trade Liberalization, Multifunctionality, and Sugar in the South Florida Landscape." *Geoforum 35* (3): 299–312.

Howard, April, and Benjamin Dangl. 2006. "Tin War in Bolivia: Conflict between Miners Leaves 17 Dead." *Upside Down World*, October 10. Web-based news site. http://upsidedownworld.org/main/bolivia-archives-31/455-tin-war-in-bolivia-conflict-between-miners-leaves-17-dead.

Jepson, Wendy. 2016. "Household Water Insecurity in the Global North: A Study of Rural and Periurban Settlements on the Texas-Mexico Border." *The Professional Geographer 68* (1): 66–81.

Kaup, Brent Z. 2010. "A Neoliberal Nationalization? The Constraints on Natural-gas-led Development in Bolivia." *Latin American Perspectives 37* (3): 123–138.

Kaup, Brent Z. 2013. *Market Justice: Political Economic Struggles in Bolivia*. New York: Cambridge University Press.

Kohl, Benjamin H., and Linda C. Farthing. 2006. *Impasse in Bolivia: Neoliberal Hegemony and Popular Resistance*. London: Zed Books.

Kohl, Benjamin H., and Linda C. Farthing. 2012. "Material Constraint to Popular Imaginaries: The Extractive Economy and Resource Nationalism in Bolivia." *Political Geography 31* (4): 225–235.

Lemos, Maria Carmen, and Arun Agrawal. 2006. "Environmental Governance." *Annual Review of Environment and Resources 31* (1): 297–325.

López Canelas, Elizabeth, Ángela C. Cuenca, Sempértegui, Silvana Lafuente Tito, Emilio R. Madrid Lara, and Patricia Molina Carpio. 2010. *El Costo Ecológico de la Política Minera en Huanuni y Bolívar*. La Paz: PIEB.

Mansfield, Becky. 2004. "Rules of Privatization: Contradictions in Neoliberal Regulation of North Pacific Fisheries." *Annals of the Association of American Geographers 94* (3): 565–584.

Marston, Andrea, and Thomas Perreault. 2017. "Consent, Coercion and *Cooperativismo*: Mining and Environmental Governance in Bolivia." *Environment and Planning A 49* (2): 252–272.

McNeish, John-Andrew. 2015. "Latin America Transformed?" In *Contested Powers: The Politics of Energy and Development in Latin America*, edited by John-Andrew McNeish, Axel Borchgrevink, and Owen Logan, 254–290. London: Zed Books.

Mirosa, Oriol, and Leila M. Harris. 2012. "Human Right to Water: Contemporary Challenges and Contours of a Global Debate." *Antipode 44* (3): 932–949.

Peck, Jamie, Nik Theodore, and Neil Brenner. 2010. "Postneoliberalism and its Malcontents." *Progress in Human Geography 41* (1): 94–116.

Perreault, Thomas. 2005. State Restructuring and the Scale Politics of Rural Water Governance in Bolivia." *Environment and Planning A 37* (2): 263–284.

Perreault, Thomas. 2006. "From the *Guerra del Agua* to the *Guerra del Gas*: Resource Governance, Neoliberalism, and Popular Protest in Bolivia." *Antipode 38* (1): 150–172.

Perreault, Thomas. 2013a. "Dispossession by Accumulation? Mining, Water and the Nature of Enclosure on the Bolivian Altiplano." *Antipode 45* (5): 1050–1069.

Perreault, Thomas. 2013b. Nature and Nation: The Territorial Logics of Hydrocarbon Governance in Bolivia. In *Subterranean Struggles: New Geographies of Extractive Industries in Latin America*, edited by Anthony Bebbington and Jeffrey Bury, 67–90. Austin: University of Texas Press.

Perreault, Thomas. 2015. "Performing Participation: Mining, Power and the Limits of Public Consultation in Bolivia." *Journal of Latin American and Caribbean Anthropology 20* (3): 433–451.

Perreault, Thomas. 2017. "Governing from the Ground Up? Translocal Networks and the Politics of Environmental Suffering in Bolivia." In *Grassroots Responses to Industrialization*, edited by Leah Horowitz and Michael Watts, 103–125. London: Routledge.

Perreault, Thomas. Forthcoming. "The Meaning of Mining, the Memory of Water: Collective Experience as Environmental Justice." In *Water Justice*, edited by Rutgerd Boelens, Tom Perreault, and Jeron Vos. Cambridge: Cambridge University Press.

Perreault, Thomas, and Gabriela Valdivia. 2010. "Hydrocarbons, Popular Protest and National Imaginaries: Ecuador and Bolivia in Comparative Context." *Geoforum 41* (5): 689–699.

Postero, Nancy Grey. 2007. *Now We Are Citizens: Indigenous Politics in Pluricultural Bolivia.* Stanford: Stanford University Press.

Prudham, W. Scott. 2004. *Knock Wood: Nature as Commodity in Douglas Fir Country.* London: Routledge.

Radcliffe, Sarah A. 2012. "Development for a Postneoliberal Era? *Sumak kawsay*, Living Well and the Limits to Decolonisation in Ecuador." *Geoforum 43* (2): 240–249.

St. Martin, Kevin. 2005. "Disrupting Enclosure in New England Fisheries." *Capitalism, Nature, Socialism 16* (1): 63–80.

Van Cott, Donna Lee. 2007. *From Movements to Parties in Latin America: The Evolution of Ethnic Politics.* New York: Cambridge University Press.

Webber, Jeffrey R. 2011. *From Rebellion to Reform in Bolivia: Class Struggle, Indigenous Liberation, and the Politics of Evo Morales.* London: Haymarket Books.

Yates, Julian S., and Karen Bakker. 2014. "Debating the 'Post-Neoliberal Turn' in Latin America." *Progress in Human Geography 38* (1): 62–90.

Zimmerer, Karl S. 2012. "The Indigenous Andean Concept of *Kawsay*, the Politics of Knowledge and Development, and the Borderlands of Environmental Sustainability in Latin America." *Publications of the Modern Language Association of America* (PMLA) *127* (3): 600–606.

Zimmerer, Karl S. 2015. "Environmental Governance through 'Speaking Like an Indigenous State' and Respatializing Resources: Ethical Livelihood Concepts in Bolivia as Versatility or Verisimilitude?" *Geoforum 64* (August): 314–324.

# 2 Mining, criminalization, and the right to protest

## Everyday constructions of the post-neoliberal Ecuadorian state

*Emily Billo*

In June 2016, I met with a worker at the Llurimagua copper mining project located in Intag, Ecuador, a cloud forest region in the western part of the province of Imbabura. Operated jointly by the Ecuadorian state mining company ENAMI (Empresa Nacional Minera) and Chile's state company CODELCO (Corporación Nacional del Cobre), the Llurimagua concession encompasses the *campesino* community of Junín (population about 270). I met with this worker, an employee of ENAMI, in the company's newly constructed office, a two-story house situated on a hill above Junín. The community had protested the presence of private mining companies in the 1990s, and more recently challenged state plans to begin exploratory operations in the community's eco-reserve. Since the election of President Rafael Correa in 2007, the Ecuadorian state has promoted its mining industry as a new frontier of resource extraction. Changes to the state mining law, ratified in 2009, have led to the designation of certain national mining regions, including the zone of Intag, where Junín is located. These legal changes, rooted in resource nationalism and post-neoliberal governance models designed to challenge neoliberal policies and practices have also led to increasingly authoritarian technologies of governance, including the criminalization of environmental protestors (Bebbington and Humphreys Bebbington 2011; Burchardt and Dietz 2014; Radcliffe 2012). As a result, citizens are subjected to the law and categorized in particular ways, leading to social control.

Detention and imprisonment have not only become common within spaces of resource extraction, they are now a manifest dimension of governing in extractive industries. Although detention and imprisonment are almost daily occurrences in communities affected by resource extraction, they are particularly noteworthy in a country like Ecuador, which has an emblematic commitment to the commons defined in its newly ratified constitution. In Ecuador, as in many countries rooted in resource extraction, the state seems to have resolved the tension between spatial mobility and containment in favor of the latter. In this chapter, I show how criminalization governs resource access and control, concentrating power in state institutions through legal and political practices that criminalize mobility within the spheres of extraction. The new geographies of extraction produced by criminalization trace the ways in which state power manipulates landscapes and people, as well as the law, to expand state sovereignty.

Even as the benefits of post-neoliberal governing cannot be dismissed—including, for example redistributive policies, social welfare, and development—we should not let these benefits obscure the risks of returning to processes of colonial racism and exclusion. Those on the left in progressive governments who continue to point to the benefits of extraction, such as jobs and development projects, still fail to provide answers for what might become of the environment, particularly clean drinking water, over the long term (Fabricant and Gustafson 2015; Radcliffe 2012).

In 2016, during my long, meandering conversation with the worker in the Llurimagua concession, I eventually asked about anti-mining protests and the increasingly violent response by the state, criminalizing those who interfere with state-run mining projects. This employee of the Llurimagua concession suggested that anti-mining residents had been "bitten by a snake, and were full of venom," emphasizing his point by pinching me on the arm. The snake, in this case, consisted of environmental activists and the local municipal government of Cotacachi, with its firmly anti-mining mayor, Jomar Cevallos. During earlier years of anti-mining protests, residents of Intag rooted their resistance efforts in the production of an environmental identity (Davidof 2013). Yet, this worker reduced residents' identities to that of a venom that poisons their bodies, permitting their criminalization. ENAMI, argued this worker, embodies the Ecuadorian state, informing people of the benefits of the mine, the state-run company's strict environmental protections and opportunities for work, and also, he suggested, engages in "finding the remedy for the snake venom." While the Ecuadorian state claims that resource extraction is for the "good of the nation," residents in Junín continue to claim their right—however fraught—to protest social and environmental harms.

ENAMI cannot arrest residents, yet its discursive tactics suggested that it upholds the law of the state, where spaces of the state are designated as national priority projects, including Intag. ENAMI contends that those who are standing in the way of resource extraction are impeding the country's development, and furthermore they do not understand the mine's benefits (Arsel et al. 2014). In 2015, ENAMI, together with national police, violently ensured their entry into the community of Junín to begin exploratory operations by policing residents, occupying the community, and putting a community leader in jail. While policing has diminished over the last year or so, its effects remain. Today, ENAMI embodies state authority by claiming to follow the law, relying on the state's discursive categorizations that permit criminalization of environmental protestors in an attempt to seek state authority and define a national identity.

The ENAMI employee continued, asking: "Should the benefit of the country be prevented by just one person: the farmer preventing access to his property?" Post-neoliberal imaginaries of the state are constructed through the criminalization of community members who identify as environmentalists or even those who claim rights to their property, a process that serves to both marginalize and also *include* these subjects in state constructions. The daily lives of populations seem indeed to show the stress of the double processes of inclusion and exclusion. This is what appears through an institutional ethnography of the effects of state responses to mining protests in Intag.

There are several conclusions to be inferred thus far that are the foundation of the remainder of this chapter. One is that the shifts in Ecuadorian state law, discourse, and practice that have led to criminalization are a means of containment of protests and dissent. Another is that criminalization must be disguised discursively by the state as a necessarily indispensable step toward the constitution of the commons. The last point, from an institutional–ethnographic perspective, is that practices associated with criminalization become integral to processes of subject formation. My argument demonstrates how the state is embodied at the local level, locating state power through spatial tactics of detention and imprisonment. I examine how resistance emerges from social relationships that inhere in the state, claiming and constructing new social relationships, and potentially new state imaginaries. In the conclusion, I explore the significance of these new geographies of resource extraction for post-neoliberal state formations in Ecuador.

## Spatial control over landscapes via criminalization

Resource conflicts in Latin America are about more than just conflicts over managing subterranean resources. Instead, they also represent struggles over identity production, citizenship, and the nation (see *inter alia* Bebbington and Bury 2013; Moore and Velásquez 2012; Perreault and Valdivia 2010; Sawyer 2004). Such struggles take on varying forms, and while they are often categorized as "communities versus the company," or "peasant struggles against rampant capital," they are typically more complex (Bebbington and Bury 2013). Opposition movements are also embedded in everyday life, and conflict often intensifies in the face of increased extraction. In the post-neoliberal Ecuadorian state, criminalization has become the state response to the tensions associated with extractive industries, leading to new forms of everyday violence associated with the post-neoliberal model.

Scholarship has alluded to the increasingly authoritarian tactics of the post-neoliberal Ecuadorian state, including criminalization, evident in places of resource extraction and other spaces deemed to be in the national interest of the state (Bebbington and Humphreys Bebbington 2011; Burchardt and Dietz 2014; Radcliffe 2012). These tactics necessitate the identification of people who oppose state practice and policy, producing detainment and detention, and concomitant socio-spatial and socio-ecological reordering at the point of extraction. Therefore, the labeling and subsequent criminalization of subjects who identify as "environmentalists" suggests simultaneous enforcement of resource extraction and reinforcement of activism and action grounded in environmental rights and protection. Labeling and excluding groups of people demonstrates exceptionalism, argues Mountz (2010). Drawing on Agamben (1998, 2005), Mountz (2010, xxviii) writes "exceptionalism is marked by paradoxical inclusions and exclusions ... along the margins of sovereign territory." Categorization is productive of particular identities, reproducing the power of the state through inclusion and exclusion (Mountz, 2010). I demonstrate that, in Ecuador, criminalization becomes a technology of governance that enables the state to meet its expanding extractive frontier and state sovereignty.

Political geographers have examined mobility and containment through detention and imprisonment, often in the context of immigration (see *inter alia* Hiemstra 2010; Martin and Mitchelson 2009; Mountz 2010; Mountz and Hiemstra 2014; Martin and Mitchelson 2009). In these cases, states operate to exclude immigrants through legal processes and spatial tactics. In Ecuador, similar techniques are employed to mark those who claim an environmental identity, where detainment produced through criminalization operates to "exclude" those who are citizens of Ecuador. As scholars (Bebbington 2009; Bebbington and Humphreys Bebbington 2011; Fabricant and Gustafson 2015; Moore and Velásquez 2012) have documented, in 2008 President Correa began referring to activists and environmentalists as "terrorists" and "dissidents," accusing them of serving the interests of multinationals, and thus perpetuating the neoliberal project. Local leaders were said to be from the city, planting false ideas about the impacts of mining in affected communities (Moore and Velásquez 2012). Moreover, these accusations targeted those who were merely seeking more information about resource extraction (see Bebbington 2009; Warnaars 2013). Even as people who identify as environmentalists demonstrate a range of concerns regarding extraction, they are all homogeneously produced via the state's process of categorization, justifying their detention and detainment. The state has implemented certain decrees, including *Decreto 16*, which forces national non-governmental organizations (NGOs) to obtain legal status to work in Ecuador and international NGOs to undergo screening to operate in the country. As a result, Correa has also temporarily and permanently closed prominent environmental organizations operating in the country (Billo and Zukowski 2015).[1] The decree seeks to undermine freedom of speech and increase government control over the groups' work. Thus, protests over resource extraction are marked by a *suspension* of the law for those who disrupt resource extraction, dismissing their rights as citizens while at the same time undermining advocacy and activist support by NGOs through the *enforcement* of legal decrees.

Scholarship that has focused on detention and detainment through criminalization emphasizes new spatialities of the state, including expanding sovereign power through new constructions of state power (see Martin and Mitchelson 2009). Sovereign power expands through the routinization of detention, through a process of "securing insecurity," facilitating subject formation (see Linneman, Wall, and Green 2014, 17). Taking an off-center view of the state, or one that uncovers the discourses, practices, and perform- ances of the state that operate to legitimate claims to rule, this project aims "to draw attention to the processes that generate different kinds of political subjects, and different kinds of political subjection" (Krupa and Nugent 2015, 5).

Following Mountz (2003, 2010), the state is an everyday social construction. Thus, the work of employees of ENAMI, for example, includes daily nation-building exercises. Employees operationalize certain technologies of governance, such as criminalization of environmentalists, and demonstrate how the power of detention and the classification of environmentalists as criminals extend into daily life. Detention by governments is an ad hoc process that necessitates reworking legal geographies, with significant impacts on those detained. Detainees are often held ambiguously, with no clear or legal determinacy. Thus, the space of extraction becomes a quasi-legal space, suspending the law for certain subject–citizens while enforcing it for others. There is no single geography of detention, but spatial tactics vary (Martin and Mitchelson, 2009). Yet, even when criminalization of environmentalists is premised on exclusion, identity production is still grounded in an inclusionary process of state-building, influencing the social relationships of state institutions. Those who are criminalized are actually "active constituents in the fabrication and maintenance of state power" where this sort of legitimate violence is most often exercised against those already dispossessed (Linneman, Wall, and Green 2014, 508).

In the present context, the state and practices of the state have become essential dimensions of everyday relationships not only between state and civil society but also between members of civil society and population groups at large. There seem to be no exceptions made even for NGOs and activists who profess resistance to the practices of the state. This at least is what I concluded from two month-long trips to Ecuador and the community of Junín in 2015 and 2016.

## Changes to state law that permit criminalization

The state has always held the rights to sub-surface resources in Ecuador, and struggles over those resources have led to debates about national belonging and sovereignty (Billo 2015; Perreault and Valdivia 2010; Sawyer 2004; Valdivia 2008). When neoliberal policies have led to private investment in resource extraction, they have also challenged state power and legitimacy, "provoking a crisis of representation and accountability" (Billo 2015; Sawyer 2004, 15). Therefore, in 2008, following President Correa's election, Ecuador's newly formed Constituent Assembly ratified a constitution that included granting rights to nature and discourses of *buen vivir (sumak kawsay* in Kichwa), translated as "living well" and in harmony with nature. The new constitution represented a rhetorical shift in Ecuador's development policies and practices, establishing more equitable and "ecologically sustainable" distribution of resources, ushering in an era often referred to as post-neoliberalism (Escobar 2010; Macdonald and Ruckert 2009; Radcliffe 2012, 241; Walsh 2010; Yates and Bakker 2014).

The Constituent Assembly also drafted a mining mandate that returned privately held concessions to the state, especially those in headwaters, ecological reserves, and where multinational companies had failed to conduct proper community consultation. Local activists supported this step toward a country free of mining. Yet, following concerns raised by Canadian mining companies with investments in Ecuador, Correa framed a new mining dialogue together with Canadian multinationals, rooted in expanding national sovereignty through state-operated resource extraction (Moore and Velásquez 2012). Thus, when President Correa passed the 2009 Mining Law, it was grounded in a more influential role for the state in resource extraction, or a neo-extractivist model of development (Gudynas 2009).

The law established a new Ministry for Non-Renewable Resources as well as the Agency for Mining Control and Regulation and the state-run mining company, ENAMI (Arsel et al. 2014; Moore and Velásquez, 2012). Furthermore, the law captured a greater proportion of rents from mining to finance state development and underwrite spending in social sectors in health and education (Dávalos 2013; Gudynas 2011; Pérez Guartambel and Solíz Torres 2014).

The 2009 Mining Law established concessionary property rights that are superimposed on local property regimes. Where a mining concession is established by the state, these interests take precedence over property rights already legally recognized by local *campesinos* or indigenous communities (Moore and Velásquez 2012). At the same time, the constitution had firmly established rights for indigenous peoples as well as prior and informed consent for any activities on their territories, preventing displacement and dispossession. The 2009 Mining Law declared the state to be the responsible institution for ensuring this consent and managing any social or environmental complaints. Yet, even when local people have the right to be consulted about a mining project in their territory, they do not necessarily have the right to prevent such a project from being implemented. The state retains the ultimate authority to determine whether a mining project can move forward (Arsel et al. 2014; Radcliffe 2012; Shade 2015). Davidof (2013) argues that the mining law essentially "neutralizes" many of the key articles in the 2008 constitution, "effectively criminalizing" any disruption that would prevent extraction (Davidof 2013, 491).

While scholars (see Bebbington and Humphreys Bebbington 2010; Fabricant and Gustafson 2015; Gudynas 2009; Moore and Velásquez 2012, 2013) have focused on the ways in which these legal changes promoted a new "twenty-first century socialism" at the expense of activists, this chapter focuses on how criminalization of environmental protests informs state institutions, producing new geographies of extraction and shaping everyday lives.

## Protest, environmental identities, and the Llurimagua concession

Resistance to mining produced Intag as a site of contentious politics, demonstrating the struggles, contradictions, and social processes that informed everyday life, which, borrowing Mountz's (2010, xxix) language marks Inteños as "simultaneously internal and external to sovereign territory". A relatively remote, marginal region, Intag became a central location for the production of identity and citizenship claims following the election of President Correa. What was once a region rooted in an environmental identity and ecological commitments that were critical of mining became a region where the state could deny residents access to their own community, criminalizing their identities.

State discourse has consistently designated Intag as a mining zone as opposed to one rooted in agriculture or biodiversity (Davidof 2013). Studies suggest that up to 72 million tons of copper could be extracted from the Llurimagua concession, accessible only via an open-pit copper mine. A subsidiary of the Japanese Mitsubishi Corporation, Bishi Metals began exploration in the mining concession in the 1990s, funded in part by the World Bank (Kuecker 2007). Coordinated resistance efforts were rooted in knowledge of the negative impacts of the mine. Support for community efforts came from a national environmental organization, Acción Ecológica, as well as the local environmental organization Defensa y Conservación Ecológica de Intag (DECOIN), established in 1995. DECOIN and Acción Ecológica's insertion into the World Bank's development plans was successful, and Bishi Metals and the World Bank pulled out of the project (Davidof 2014; Kuecker 2007).

In 2004 the former Canadian-listed mining company Ascendant Copper expressed interest in the Llurimagua project. However, the *cantón* of Cotacachi declared that the region was a *cantón ecológico* (ecological county), banning mining and instead promoting environmentally sustainable projects (shade-grown coffee cooperatives, sugar production, and eco-tourism, for example) (Bebbington et al. 2007). Control over livelihoods and income at the local level informed an environmental identity and the formation of environmental subjects (Agrawal 2005). This environmental identity, together with a violent assault on the residents of Junín by paramilitaries associated with Ascendant Copper in 2006, fueled a broad-based resistance movement. Ascendant Copper was forced out of the region in 2008 (Davidof 2014).

Following the election of President Correa, Intag was designated as a national strategic zone because of its potential for copper extraction. The arrival of the state company, ENAMI, and Chile's CODELCO in 2014 was met with community resistance, employing the tactics of earlier efforts, such as roadblocks. These efforts proved ineffective when confronted by national police. Today soil sampling is underway throughout the concession, part of the exploratory phase of the project. ENAMI and CODELCO have signed an agreement for this initial phase, and if the region is deemed productive by CODELCO the state will negotiate an additional agreement for the extraction phase. About 60 police maintain an active presence in the region to ensure that operations continue without interruption.

The shift in extractive policies following President Correa's election led to spatial and territorial reordering rooted in post-neoliberal claims that resource extraction would bring development to the region. The state's responses to Intag's environmental identity were grounded in the violence of criminalization, whereby local claims to land were made illegal and the categorization of environmentalists labeled local activists as criminals, terrorists, and dissidents. Efforts to categorize and criminalize environmentalists informed state rhetoric that resource extraction was an important step in promoting the commons. Criminalization introduced certain tactics of spatial containment, with concomitant effects on residents of Junín, explored in the following sections of this chapter.

## Containment through imprisonment

The subsequent anecdote illustrates how routine arrests or assaults can result in processes of self-surveillance or state control over spatial mobility (see Mountz 2010; Hiemstra 2008). On April 11, 2014 Javier Ramírez, president of Junín, was arrested. The state accused Ramírez of harming ENAMI employees who were beginning copper mining explorations. Despite local and international claims that Ramírez was not present during protests against the state mining company, he was imprisoned for 10 months. After Ramírez's arrest, 300 national police were sent to the region to "protect" ENAMI and Chile's CODELCO.

Javier Ramírez was imprisoned based on spurious claims by the Ecuadorian state that he had participated in and injured a state mining company employee by throwing a rock during anti-mining protests in Junín. Pulled off of a bus on his way to Quito by Ecuadorian national police with no arrest warrant, Javier was ultimately imprisoned in Ibarra, the capital of Imbabura province. In June 2016, at home in Junín, Javier told me that he has to "take more precaution" with the people with whom he speaks, people he does not know, and those who are employed by the state. Imprisonment has the effect of producing "would-be criminals," which can lead to "self-policing," including limiting one's visibility and mobility (Bigo 2007, 4; Hiemstra 2010; Martin and Mitchelson 2009).

The "legal indeterminacy" of Javier's arrest initiated a process of political silencing of Javier and the anti-mining movement, including a means of containing protests (Martin and Mitchelson 2009, 465). Amnesty International along with a national human rights organization brought attention to Javier's case. Despite the claims of several witnesses, including Javier's employer, a Belgian doctor who owned a coffee farm in Junín and who publicly declared that Javier was home recovering from a motorcycle accident, the state refused to incorporate these testimonies into its case against Javier. In the 10 months Javier spent in jail, no charges were brought against him. Javier's arrest demonstrates how the space of Junín has become a site where detention is normalized and routinized.

## Imprisonment as expanding social control

As Martin and Mitchelson (2009) note, even as criminalization and imprisonment may remove those detained from public view, we must still consider the effects of detention on family members who bear the burden of imprisonment. Alejandra Santanilla Ortíz of *Instituto de Estudios Ecuatorianos* told me in an interview in Quito in 2016 that criminalization is about much more than criminalizing an individual. The state's actions strategically target families of activists and leaders who "take years to form." The Ramírez family has a long history of activism in Intag. Javier's father was killed by a neighbor in earlier struggles linked to mining activity; he refused to open the family's land to mining companies (see Davidof 2013). As I chatted with residents of Junín, including members of the Ramírez family, I learned they are all firmly engaged in the anti-mining movement. Family members all told me it was something they just knew how to do, rooted in how they lived their lives—on the land, engaged in agriculture. This way of life is what they wanted to hold onto for their children. Javier was president of the community at the time of his arrest, and the false accusations by the state about his participation in the anti-mining protest demonstrate a much more strategic arrest on the part of the state. Furthermore, Javier was not the only member of his family accused of anti-mining activities. During the period of Javier's arrest, his brother, Victor Hugo, remained hidden in Intag, eluding state officials.

In the case of Intag, the impacts of criminalization on families are typically gendered, as women often take on the invisible role of providing support for imprisoned family members. As Javier served his sentence, his family, including his wife, mother, and children, lived through the uncertainty regarding the length of his sentence and its outcome. As Javier described it: "it was desperation for my family . . . I had never left [Junín] before—not for work, not for fun." His wife and mother spent several days at a time in Ibarra, waiting for word on Javier's case and ensuring they were present on all visiting days. Javier's wife suffered physical pain and illness, the result of stress from supporting Javier while he was imprisoned. These physical illnesses continue today. This left Javier's children at home, as an aunt took care of them, so that their lives might continue without further disruption.

Javier's children also suffered from the effects of criminalization. There were long gaps when his three children did not attend school, falling behind in their academic work. When they did attend school, they endured the harassment of students and teachers from the pro-mining county of García Moreno (see also Warnaars 2012). Recently, these pressures caused Javier's oldest son to leave high school before graduating. His shoulder-length hair was said to violate policies of the state-run public school, and he refused to cut it.

Criminalization demonstrates a process of surveillance, marking some as permitted and others as criminal. Its effects extend beyond the rule of law into everyday activities, such as going to school. Some residents of Junín have received medical assistance to help them recover from the physical violence of the state and also to address the psychosocial impacts of policing (Acción Ecológica 2015). In addition to these effects, state control continues via a process of categorizing residents as justifiably criminal, a policy I explore in the next section.

## Subject formation as spatial containment

The kind of visibility criminalization inflicts on residents, such as imprisonment, monitoring family interactions while at home, policing residents as they walk through their communities, and demanding to see identification papers, suggests a form of spatial control (Foucault 1991, 1995; Mountz 2010; Secor 2007). The internalization of state responses to mining—which are inevitably tied to questions of state power—can inform subjects' environmental and political identities. A focus on identity and subject formation can reveal the structures or systems of power that underlie state policy, setting the conditions for social control (Nightingale and Ojha 2013).

On a trip to Junín's eco-reserve in 2015, where ENAMI was beginning its exploratory operations, we climbed to the mining access point. The community-owned reserve is popular for eco-tourism, and includes dramatic waterfalls. The actions of ENAMI in the community reserve demonstrate material evidence of the undermining of community members' environmental identities. As we reached a junction from which one trail led to the waterfalls while the other led to the exploratory mining operations, two workers blocked our way. "Nobody passes without company permission," the workers informed us. The company was refusing entry to community members, activists, and researchers, citing a lack of safety gear, even though on this trip we had come with vests and hardhats. Clearly, these rules were applied arbitrarily. Employees of ENAMI passed by without the requisite safety gear, leading a mule train up the trail.

We were forced to exit the reserve, although we took our time, stopping by the waterfalls on our way down. Our slow descent down the trail alerted the company and national police. When we finally did exit, three members of the national police were waiting for us. They were nervous that we had tried to enter the mining operations via a circuitous route. In response to our questions concerning their presence, they answered, "We're here to keep the peace between the *ecologistas* and *mineros*." This statement suggested a clear categorization of residents into two groups, one that was permitted to enter the reserve, now transformed into the space of the concession, the other to be excluded. Our community guide urged us to continue, to not interact with the police. He was nervous about our questions, revealing the fear invoked by this subjection by the state. An additional police presence awaited us at the next trail junction, ensuring that we would leave the reserve. Yet, although those categorized as *ecologistas* are prevented from accessing the space of the reserve, *mineros* are not limited in their spatial mobility. These inequalities are forged on the bodies of anti-mining residents and reflected in their uneven access to community spaces.

Those identified as anti-mining were treated more harshly than those who were not. Policing and criminalization has divided families and neighbors, and participant observation and interviews have suggested that pro-mining and anti-mining groups rarely speak to or interact with one another. In a small community like Junín, these divisions have dramatic effects on people's daily lives, driving people to keep to themselves and end relationships with community and family members. These divisions become starker when the state suggests criminalization is necessary if resource extraction is to be shared in common.

## Expanding state power via criminalization

In this section I examine the ways in which criminalization is used to disguise the violence of state practices, and I also demonstrate its essential role in producing resource access and control. Interview respondents told me that lack of work for young people remains a challenge in Intag (see also Bebbington et al. 2007; Davidof 2013). Yet, as Davidof (2013) points out, this discourse that references the lack of economic opportunities replaces and disguises earlier concerns rooted in the social and environmental impacts of mining. Because state employees are familiar with poverty as well as the lack of educational opportunities and employment in Intag, the call of ENAMI and a state rooted in resource extraction is desirable and prestigious. Indeed, some residents who had left the region looking for jobs returned when news of ENAMI's presence materialized. ENAMI embodies the state and sovereignty, and new employees, like those described here, develop new worldviews as they are socialized into the company. The social and environmental consequences of mining become reinterpreted as poverty that mining could mitigate. Residents who call attention to these social and environmental impacts, or *ecologistas*, are standing in the way of progress and modernization. Therefore, employment opportunities offered by ENAMI are also strategic, rooted in the social relations of the community.

ENAMI employs young women, 18–20 years old, who work as *socializadoras*, visiting houses in communities to inform residents of the implications of the mine. Older women are also employed to cook, provide housing, and wash clothes for miners, while men are employed in the more physically demanding work of exploration in the eco-reserve. Employment is offered to community members to convince them of the economic opportunities the mine provides. Most say they will save money for their own or their children's education, or to buy a house in a nearby city.

Women employed by ENAMI were extremely cautious when speaking with me, and eventually outright refused to do so. They also refused to allow me to take any notes or record our brief conversations. State discourse remains coherent by threatening employees with unemployment (and potentially criminalization) for any slippage of information (see also Belcher and Martin 2013). ENAMI constructed an office building in Junín, which sits on a hill above the community, marked by a sign and now lit up at night. The company literally watches over the community. In 2015 a resident of Junín told me, "They continue watching us. ENAMI is like an observer," referring to this constant presence of the company in the community.

Although ENAMI employees have roots in Intag and experience the day-to-day violence of criminalization, I found almost no one who was willing to reflect on the contradictions in their identity as both a resident and an employee of the state. The fear of policing, together with the perceived economic wealth that mining brings, has structured social relationships at the local level. Indeed, it is quite difficult to meet with employees of ENAMI (or other state officials). They are extremely nervous about being identified in media or academic publications, and are loath to produce any weakness in the state's pro-mining discourse. For example, when attempting to speak to the *socializadoras* in their office in Junín, I was referred to their boss, the director of community relations. The *socializadoras* told me that he would be there "tomorrow," and although I returned to the office every day, I was always told to come back the next day. As Secor (2007) observed in her research on the Turkish state, this constant waiting and submission is also an exercise in the state's power over my role as researcher. Contrary to what Mountz (2003) reports in her study of the Canadian state, these ENAMI employees maintain the coherence of the state even as they interact with residents and researchers who raise questions and concerns, and their own identities are rooted in the contentious region of Intag. Institutional closure is an expression of state power (Belcher and Martin 2013).

## Productive power and resistance

Environmentalists and activists portray a state that is out of control, and thus state policy must emerge to manage protests. The state needed to construct particular identities to exercise control—the power of the state is visible through discourse that categorizes residents as criminal—yet the construction of identities is never as neat as the state imagines. Moreover, state affect often produces an outcome that challenges the intended effects of state policy, where resistance emerges from within (Mountz 2010; Sawyer 2004). In Intag, post-neoliberal policies have continued to produce resistance.

While Javier Ramírez was in prison, the company tried to garner support among residents. It established a state-backed *cabildo* (local council government) even as anti-mining residents established their own *cabildo*. The state attempted to divide community members, seeking more support for company operations, extending into the local level, expanding its reach, pressuring and threatening residents to vote to support the company-backed *cabildo*. This dual leadership system remained in place until December 2015. Elections are held each year in the community, and Javier ran again for community president—challenged by a company-backed resident. Javier won the election, but this outcome was questioned by the state's *teniente politico* (the political lieutenant of the parish council—the lowest level of local government). A second election was held, this time with secret ballots. As is typical, pressure was put on residents employed by ENAMI to vote for the state-backed candidate or risk losing their jobs. In Junín, in December 2015, Javier won the election a second time, ensuring that he was the new community president. Despite state efforts to criminalize the candidate, support for Javier did not wane. Criminalization was not enough to mark him as an ineffective candidate, and while it temporarily removed him from the community it did not diminish his leadership role.

Since Javier's reelection, resistance to state practices has continued, but tactics have changed. Promised infrastructure, including paved roads, a school, and a hospital, are nowhere to be found in Intag. These promises were part of the state's initial appeal to residents as it started mining operations, suggesting that the mine would bring progress, modernization, and development to the region. Now resistance hinges on these failed promises, criticizing the narrative of the strong, progressive state. The material shortcomings experienced by residents shifts social relationships, offering a counter-narrative to the one of progress and modernization promoted by the central state.

Activists demonstrate their knowledge of state authority, which helps them see through and around state discourse and practice. Despite failures to deliver infrastructure, ENAMI officials claimed that they never promised these improvements to community members. Instead, they would begin infrastructural improvements only when the mine proved to be successful—potentially in 3 or 4 years. Residents were quick to interpret this change in the state's narrative as a form of tricking people into allowing the company to enter and begin exploration. It also uncovers another illusion—Junín would have to be relocated if the mine proved successful. Why, residents asked, would the government ever build a school, hospital, or improve roads in a community that would soon cease to exist? The most critical of residents already recognized the shortcomings of this discourse when ENAMI first entered the community, realizing that the offer of infrastructural improvements by ENAMI would never come to fruition. For those who were hoping they were justified in believing otherwise, or perhaps were told to believe otherwise through the threat of criminalization, they too have now started to see through these illusions of modernity and progress, joining the resistance movement. As people begin to point out these shortcomings in state promises and clamor for infrastructural improvements, they also argue that the state is supposed to operate differently. Protests demonstrate the incoherence of state discourse and practice.

Even as the state criminalizes residents, this does not always have the intended effect. Despite Javier's wariness about speaking with others following his arrest, he still sees himself as a citizen of Ecuador, but one invested in building a better life for himself and his family—rooted in the environment, agriculture, and the land. For Javier, the environment represents the production of novel social relationships. Resource extraction is not going to allow people to live, he said in an interview in 2016. Instead, he argued, Ecuadorians need to look for and understand new relationships to their environment. Backed by local, national, and transnational environmental activists and organizations as well as the municipal government of Cotacachi, Javier continued to produce alternative narratives of the state, making visible those practices, people, and spaces that are concealed by dominant state discourses.

In mid-June 2016, Acción Ecológica, together with local activists from around the country affected by mining, organized a protest in Quito calling on the Ministry of Mines (established in 2015) to halt all mining operations—in short, seeking a mining moratorium. The government had just defined new concessions in the country, including 14 in the Intag region. The concessions were established with no prior consultation of residents in the region. The national government would "auction" these concessions to the highest bidder, whether a state or private entity or an individual. Quickly, exploratory operations might begin in the entire region of Intag. This auction was likely related to the country's insolvency, demonstrating the rootedness of resource nationalism in political economy (see Perreault and Valdivia 2010). The price of petroleum had dropped over the past several years and the price of copper remained quite low. ENAMI was laying off workers in the Llurimagua concession. Indeed, global commodity prices offered some hope that the state would begin looking for development alternatives instead of pushing forward with mining. Yet, the immediate response—auctioning off additional concessions—seemed to suggest otherwise. The potential expansion of mining concessions in Ecuador, together with tactics that highlighted shortcomings of state narratives of progress and modernization rooted in resource extraction, produces contradictions that open new spaces in which to rethink configurations of the nation-state.

## Conclusion

Places of resource extraction represent national priority projects, where state law is both suspended for environmentalists who might disrupt resource extraction and enforced for advocacy and activist environmental and human rights organizations. These legal ambiguities and associated forms of spatial control via detention and detainment and their effects suggest new geographies of resource extraction in Ecuador. The discursive representation of environmentalists as "terrorists" or "dissidents" in national discourse and media representation, together with policies that withdraw support for advocacy organizations, undermines certain citizens, even as state policy does not explicitly advocate for criminalization. Law and policy are written into wider state and transnational networks, and this means that, as resource extraction and protests expand across the region, criminalization produces new forms of state power. The routinization of criminalization and policing suggests expanding sovereignty via increased insecurity, making the production of criminalized subjects essential to state formation. In other words, the resulting spatial transformations at the point of extraction lend additional insight into the post-neoliberal, developmental Ecuadorian state.

Employing an institutional ethnography and an embodied analysis of the post-neoliberal Ecuadorian state, this chapter provides a critical investigation into social relationships that are involved in criminalization and articulated through discourses, practices, and performances of subjects in the space of Intag. Through vignettes that examine imprisonment, spatial containment, and social control via subject formation, these spatial configurations emphasize how residents of Junín have become "active constituents" in the production of state power and expanding sovereignty (Linneman, Wall, and Green 2014, 508). These geographies restructure socio-ecological and spatial relationships at the point of extraction, demonstrating the ways in which violent forms of spatial containment are exercised against those who are already marginalized.

This approach underscores, in particular, the ambivalence and tensions that result from the contradiction between the policies and practices outlined in the 2008 constitution that was designed to protect the rights of the environment and citizens as well as the daily lives of residents in places of extraction. The arbitrary and uneven criminalization of citizens, together with ENAMI's claim that resource extraction operates for the "good of the nation," recalls the violent exclusionary practices of neoliberalism. Criminalization is an essential step in the enforcement of resource extraction for the common good.

Embodied processes of self-surveillance, detention, and criminalization expand state presence in spaces of resource extraction. The self-policing and spatial containment of subjects in the Llurimagua concession demonstrate the potential for criminalization to block the emergence of alternative forms of development and progressive change. Yet, ethnography also offers opportunities for changing dominant narratives and practices, exploring spaces for resistance. Intag illustrates that organizing at local, regional, national, and transnational scales will continue to pressure governments to abandon tactics of criminalization and subjectification, as local residents see through processes of categorization, reclaiming an environmental identity by supporting those who desire development on their own terms.

Even as I argue that the space of Intag is one of exception, the practices of criminalization are by no means an exception in Ecuador or Latin America. Intag is but one case in Ecuador and the region—a case that signals a growing trend toward marginalization, violence, and dispossession under post-neoliberal governments. Additional research must investigate the ways in which processes of classification and categorization also discriminate based on race, gender, sexual identity, and so on. Similarly, further research should explore the ways in which these patterns are embedded in historical processes of colonial exclusions, or post-colonial exclusions perpetuated in a post-neoliberal process (see Radcliffe 2012). The contradictory processes of state formation in Ecuador have led to co-optation of social movements and increased struggles over resource extraction (see Walsh 2010) and, in turn, laws are both suspended and enforced to facilitate concentration of state power, through which criminalization produces new patterns of indeterminacies, new institutional arrangements, and new geographies of resource extraction in Ecuador.

## Note

1   In January 2017, President Correa's government attempted to stop the prominent national environmental organization Acción Ecológica for the second time, claiming that the organization had incited protests related to a mining project that continues to affect indigenous Shuar populations in the southern Amazon region. The Ministry of Environment ultimately overturned the recommendation to shut down the organization's operations.

## References

Acción Ecológica. 2015. *Íntag: Una Sociedad que la Violencia no Puede Minar*. Quito, Íntag, Ecuador: El Chasqui Ediciones.

Agamben, Giorgio. 1998. *Homer Sacer: Sovereign Power and Bare Life*. Stanford: Stanford University Press.

Agamben, Giorgio. 2005. *States of Exception*. Chicago: University of Chicago Press.

Agrawal, Arun. 2005. *Environmentality: Technologies of Government and the Making of Subjects*. Durham, NC: Duke University Press.

Arsel, Murat, Carlos Mena, Lorenzo Pellegrini, and Isabella Radhuber. 2014. "Property Rights, Nationalisation and Extractive Industries in Bolivia and Ecuador." In *Conflicts over Natural Resources in the Global South—Conceptual Approaches*, edited by Maarten Bavinck, Lorenzo Pellegrini, and Erik Mostert, 109–128. Leiden, Netherlands: CRC Press.

Bebbington, Anthony. 2009. "The New Extraction: Rewriting the Political Ecology of the Andes?" *NACLA*, September/October: 12–40. doi: 10.1080/10714839.2009. 11722221.

Bebbington, Anthony, Jeffrey Bury, Denise Humphreys Bebbington, Jeannet Lingan, Juan Pablo Muñoz, and Martin Scurrah. 2007. "Moviemenientos Sociales, Lazos Transnacionales y Desarollo Territorial Rural en Zonas de Influencia Minera: Cajamarca-Peru y Cotacachi-Ecuador." In *Minería, movimientos sociales y respuestas campesinas. Una ecología política de transformaciones territoriales*, edited by Anthony Bebbington, 163–183. Lima: Instituto de Estudios Peruanos.

Bebbington, Anthony, and Denise Humphreys Bebbington. 2011. "An Andean Avatar: Post-Neoliberal and Neoliberal Strategies for Securing the Unobtainable." *New Political Economy 16* (1), 131–145.

Bebbington, Anthony, and Jeffrey Bury. 2013. "Political Ecologies of the Subsoil." In *Subterranean Struggles: New Dynamics of Mining, Oil, and Gas in Latin America*, edited by Anthony Bebbington and Jeffrey Bury, 1–25. Austin: University of Texas Press.

Belcher, Oliver, and Lauren L. Martin. 2013. "Ethnographies of Closed Doors: Conceptualising Openness and Closure in US Immigration and Military Institutions." *Area 45* (4): 403–410.

Bigo, Didier. 2007. "Detention of Foreigners, States of Exception, and the Social Practices of Control of the Banopticon." In *Borderscapes: Hidden Geographies and Politics at Territory's Edge*, edited by Prem Kumar Rajaram and Carl Grundy-Warr, 3–33. Minneapolis: University of Minnesota Press.

Billo, Emily. 2015. "Sovereignty and Subterranean Resources: An Institutional Ethnography of Repsol's Corporate Social Responsibility Programs in Ecuador." *Geoforum 59*: 268–277.

Billo, Emily, and Isaiah Zukowski. 2015. "Criminals or Citizens? Mining and Citizen Protest in Correa's Ecuador." *NACLA*, November 2. https://nacla.org/news/2015/11/02/criminals-or-citizens-mining-and-citizen-protest-correa's-ecuador.

Burchardt, Hans Jürgen, and Kristina Dietz. 2014. "(Neo)-extractivism: A New Challenge for Development Theory from Latin America." *Third World Quarterly 35* (3): 468–486.

Dávalos, Pablo. 2013. "'No Podemos Ser Mendigos Sentados en un Saco de Oro': Las Falacias del Discurso Extractivista. In *El Correísmo al Desnudo*, edited by Alberto Acosta, 190–215. Quito: Montecristi Vive.

Davidof, Veronica. 2013. "Mining versus Oil Extraction: Divergent and Differentiated Environmental Subjectivities in 'Post-Neoliberal' Ecuador." *Journal of Latin American and Caribbean Anthropology 18* (3): 485–504.

Davidof, Veronica. 2014. "Land, Copper, Flora: Dominant Materialities and the Making of Ecuadorian Resource Environments." *Anthropological Quarterly 87* (1): 31–58.

Escobar, Arturo. 2010. "Latin America at the Crossroads: Alternative Modernizations, Post-Liberalism, or Post-Development?" *Cultural Studies 24* (1): 1–65.

Fabricant, Nicole, and Bret Gustafson. 2015. "Moving Beyond the Extractivism Debate, Imagining New Social Economies." *NACLA Report on the Americas 47* (4). https://nacla.org/article/moving-beyond-extractivism-debate-imagining-new-social-economies.

Foucault, Michel. 1991. "Governmentality." In *The Foucault Effect: Studies in Governmentality*, edited by Graham Burchell, Colin Gordon, and Peter Miller, 87–104. Chicago: University of Chicago Press.

Foucault, Michel. 1995. *Discipline and Punish: The Birth of the Prison*. New York: Random House.

Gudynas, Eduardo. 2009. "Diez Tesis Urgentes sobre el Nuevo Extractivismo: Contextos y Demandas bajo el Progresismo Sudamericano Actual." In *Extractivismo, Política y Sociedad*, 187–225. Quito: CLAES, CAAP.

Gudynas, Eduardo. 2011. "Alcances y Contenidos de las Transiciones al Post-Extractivismo." *Revista Ecuador Debate 82*: 61–79.

Hiemstra, Nancy. 2010. "Immigrant Illegality as Neoliberal Governmentality in Leadville, Colorado." *Antipode 42* (1): 74–102.

Krupa, Christopher, and David Nugent. 2015. "Off-Centered States: Rethinking State Theory through an Andean Lens." In *State Theory and Andean Politics: New Approaches to the Study of Rule*, edited by Christopher Krupa and David Nugent, 1–31. Philadelphia: University of Pennsylvania Press.

Kuecker, Glen David. 2007. "Fighting for the Forests: Grassroots Resistance to Mining in Northern Ecuador." *Latin American Perspectives 153* (34, 2): 94–107.

Linneman, Travis, Tyler Wall, and Edward Green. 2014. "The Walking Dead and Killing State: Zombification and the Normalization of Police Violence." *Theoretical Criminology 18* (4): 506–527.

Macdonald, Laura, and Arne Ruckert. 2009. *Post-Neoliberalism in the Americas*. New York: Palgrave Macmillan.

Martin, Lauren L., and Matthew L. Mitchelson 2009. "Geographies of Detention and Imprisonment: Interrogating Spatial practices of Confinement, Discipline, Law, and State Power." *Geography Compass 3* (1): 459–477.

Moore, Jennifer, and Teresa Velásquez. 2012. "Sovereignty Negotiated: Anti-Mining Movements, the State and Multinational Mining Companies under Correa's '21st

Century Socialism.' " In *Social Conflict, Economic Development and Extractive Industry: Evidence from South America*, edited by Anthony Bebbington, 112–133. London and New York: Routledge.

Moore, Jennifer, and Teresa Velásquez. 2013. "Water for Gold: Confronting State and Corporate Mining Discourses in Azuay, Ecuador." In *Subterranean Struggles: New Dynamics of Mining, Oil, and Gas in Latin America*, edited by Anthony Bebbington and Jeffrey Bury, 119–148. Austin: University of Texas Press.

Mountz, Alison. 2003. "Embodying the Nation-State: Canada's Response to Human Smuggling." *Political Geography 23* (3): 323–345.

Mountz, Alison. 2010. *Seeking Asylum: Human Smuggling and Bureaucracy at the Border.* Minneapolis and London: University of Minnesota Press.

Mountz, Alison, and Nancy Hiemstra. 2014. "Chaos and Crisis: Dissecting the Spatiotemporal Logics of Contemporary Migrations and State Practices." *Annals of the Association of American Geographers. 104* (2): 382–390.

Nightingale, Andrea J., and Heman R. Ojha. 2013. "Rethinking Power and Authority: Symbolic Violence and Subjectivity in Nepal's Terai Forests." *Development and Change 44* (1): 29–51.

Pérez Guartambel, Carlo and Maria Fernanda Solíz Torres. 2014. "Territorio, Resistencia y Criminalización de la Protesta." In *La Restauración Conservadora*, edited by M. Aguirre, 153–166. Quito: Montecristi Vive.

Perreault, Tom, and Gabriela Valdivia. 2010. "Hydrocarbons, Popular Protest and National Imaginaries: Ecuador and Bolivia in Comparative Context." *Geoforum 41* (5): 689–699.

Radcliffe, Sarah A. 2012. "Development for a Postneoliberal Era? *Sumak kawsay*, Living Well and the Limits to Decolonisation in Ecuador." *Geoforum 43* (2): 240–249.

Sawyer, Suzana. 2004. *Crude Chronicles: Indigenous Politics, Multinational Oil, And Neoliberalism in Ecuador.* Durham, NC and London: Duke University Press.

Secor, Anna J. 2007. "Between Longing and Despair: State, Space, and Subjectivity in Turkey." *Environment and Planning D: Society and Space 25* (1): 33–52.

Shade, Lindsay. 2015. "Sustainable Development or Sacrifice Zone? Politics below the Surface in Post-Neoliberal Ecuador." *The Extractive Industries and Society 2* (4): 775–784.

Valdivia, Gabriela. 2008. "Governing Relations between People and Things: Citizenship, Territory, and the Political Economy of Petroleum in Ecuador." *Political Geography 27* (4): 456–477.

Walsh, Catherine. 2010. "Development as *Buen Vivir*: Institutional Arrangements and (De)colonial Entanglements." *Development 53* (1): 15–21.

Warnaars, Ximena S. 2012. "Why Be Poor When We Can Be Rich? Constructing Responsible Mining in El Pangui, Ecuador." *Resources Policy 37* (2): 223–232.

Warnaars, Ximena S. 2013. "Territorial Transformations in El Pangui, Ecuador: Understanding How Mining Conflict Affects Territorial Dynamics, Social Mobilization, and Daily Life." In *Subterranean Struggles: New Dynamics of Mining, Oil, and Gas in Latin America*, edited by Anthony Bebbington and Jeffrey Bury, 149–171. Austin: University of Texas Press.

Yates, Julian S., and Karen Bakker. 2014. "Debating the 'Post-Neoliberal Turn' in Latin America." *Progress in Human Geography 38* (1): 62–90.

# 3 Preserving illusions

## The rule of law and legitimacy under the Chad Pipeline Project

*Siba N. Grovogui*

This chapter was inspired by field research in the oil fields of Chad. Yet, its focus remains conceptual. It focuses on the utility of the rule of law as a metaphor of state, corporate, and global governance. It is based on circumstances and conjectures concerning the context of the Chad Oil and Pipeline Project to highlight a number of things. The first is that language, including the language of the rule of law, has multiple valences. Historically, the rule of law has been understood to establish: (1) a constitutional order in which government and the governed are bound by and subject to the law; (2) equality before the law and the protection of human rights; (3) administration of justice in a fair, expeditious, and transparent manner; and (4) the accessibility to all of constitutionally-prescribed or customary processes, procedures, and rules. It is a contention of this chapter that these understandings of the rule of law only partially hold under neoliberalism.

The second contention that this chapter makes is that the deployment of the concept of the rule of law, outside the strictures of its traditional liberal applications, creates different effects. In this sense, the rule of law is currently deployed under neoliberal institutional arrangements to imply commitments that are clearly unattainable under present state, corporate and international forms. This may be illustrated, for instance, by the emergence of the notion that property is *the* essential, actionable, and constitutionally guaranteed individual right.

My final argument is that the neoliberal redefinition of the extent and meaning of the rule of law is not happenstance. It is intended to conscript both the state and international organizations, including non-governmental organizations, into different sets of moral commitments articulated by corporations and financial capitalism. Yet, rather than elaborating on all these points, this chapter locates itself in the gap between the rule of law, an idea that has ideological potency, and its outcomes. These outcomes can be described as a set of unpredictable relationships and events that betray the central assurance of predictable relationships offered by conventional understandings of the rule of law.

Within this new legal environment that neoliberal redefinitions of the term inaugurate, the rule of law appears as a stylized diction of intent and purpose, in search of formal validity as the instrument of or means to the common good. The value and the accuracy of this universal pretense is revealed partly in the tone of the enunciations of laws that are intended to satisfy a need and, therefore, necessarily reflect an ecology of moods, desires, and dispositions. Yet, the most important dimension of this understanding of the rule of law is the condition of necessity embedded in its enunciations.

In Chad, this condition is the desire of investors and the state to ensure the uninterrupted exploration, extraction, and safe delivery of oil to the market. From this perspective, the objective of the rule of law is to bring about a normative regime tailored to a pre-ordained market ideology, what I mean here by neoliberalism. This is reflected in the structures of the instituted relationships between and among, on the one hand, the World Bank, the oil industry, and the Chadian state, and on the other, the populations of Chad, which have under these conditions been divided into the categories of the nation, the inhabitants of the oil field region, the expropriated, and contract laborers. This logic is therefore also evident in the manners in which the World Bank and the oil consortium leaned on the state to dismantle, modify, or simply erase from institutional memory anterior national, regional, cultural, and legal norms bearing on property, political and social order, and social relations.

The overall argument presented here then is that the rule of law is a conjecture, a logical construction based on incomplete information. This incompleteness is the stuff of presumptions, assumptions, postulations, and suppositions that are proper to an object. This sense of the rule of law becomes apparent when one examines the political and judicial *dispositifs* set in place for the purpose of poverty reduction. These take the form of instruments and mechanisms of investment; rules applied to contractual relations and compensation; and forms for the adjudication of disputes among contestants. From this angle then, the rule of law is neither a matter of conspiracy, nor an unvarnished display of generosity. It is a matter of design and legislation, of contentious implementation by multiple agents, and, finally, of intervention by the different organs of state, capital, and transnational organizations. It would thus be mistaken to take at face value the proposition that the rule of law, even in the context of development in Chad, was intended to effect the harmonious exploitation of oil and distribution of resources. The goal of the rule of law, here and elsewhere, was to create a sense of legitimacy and an aura of consent on the part of the populations it enumerated, in the ambition to produce an image of consensual relations between the expropriated and laborers and the state, oil companies, and the World Bank.

My goal is ultimately to peel away at the veneer of the universal goodness that the neoliberal rule of law that has been installed in Chad attempts to signify. I show that the regime of laws implemented in Chad was modulated specifically to the assumed needs of a supposed failed post-conflict state. This understanding of Chad enabled relationships that one would not ordinarily associate with the rule of law. In other words, in place of an organic connection between the will of the citizenry and the laws by which the former are governed, national and corporate administrators in Chad favored regulatory regimes that created distance and degrees of separation between the sovereign and the citizenry, administrators and the administered, and the oil industry and the populations affected by the exploitation of oil. This distancing has, in turn, given rise to publicly impenetrable zones of regulatory exception and therefore micro-sovereign privileges and immunities to different entities.

The pipeline project allows us to ask new questions about the rule of law and its complex web of economies, social orders, and modes of governance. In the context of neoliberal market ideology and modes of governing, the rule of law means immunities, privileges, liabilities and other legal and political advantages and disabilities instituted in favor of corporate, state, and international actors concerned with extracting oil and getting it to market. The products of this form of legal rule are intended to guarantee specific outcomes: (1) security and revenues for the state; (2) crude oil for the corporation; and (3) a stable political economy for the World Bank. These activities and resources are then converted into fuel for the economy and goods for the consumer. The rule of law, thus understood, compels state, corporate, village, and community-level executives, arbitrators, and enforcers to an end that may not lead to poverty reduction.

## The rule of law in historical perspective

The rule of law has a long and rich trajectory in legal and moral thought. The appeal of the idea—as opposed to its various institutional iterations—is its supposed moral foundation. This is that the rule of law not only inscribes desirable and legitimate moral and legal precepts into public life as guide to behavior. It also sets up clear procedures for the attainment of justice that remain universally accessible to all (Raz 2009).

A few implications necessarily follow from this supposition. Throughout much of the modern era the rule of law has been understood as a safeguard against arbitrary rulings against citizens in judicial proceedings in individual cases (Locke 1980; Bowen 2013; Berlin 1969). The morality of this principle is self-evident. It is predicated upon the assumption that the government exercises its authority in accordance with written laws, adopted through established constitutional procedures, by duly appointed or elected legislators. This also means that judicial institutions exist prior to litigation and judicial rules and procedures do not change in manners that are prejudicial to a supplicant based on the humors of the governor or sovereign. More recently, these points have been refined in academic and other professional circles to imply that

> all laws should be prospective, open, and clear; laws should be stable; the making of laws should be guided, open, clear, and [based on] general rules; the independence of the judiciary must be guaranteed; natural justice must be observed; courts must have reviewing power over some principles;

courts should be accessible; and the discretion of crime-preventing agencies should not be allowed to pervert the law.

(Raz 2009, 210)

These principles are by now matters of commonsense.

Despite this apparent convergence in theoretical and public discourses on the rule of law, there remains much disagreement among theorists and practitioners alike about the purpose, object and mechanisms of the rule of law. Even as most hold that the rule of law should promote civil liberties (a trend that began in the English-speaking world with the Magna Carta), there is variance among scholars and politicians over the extent of the regulatory authority of the state and, additionally, police power. For instance, one need only recall the legislative and judicial developments in the United States following the events of September 11, 2001. These developments demonstrated great variation in the perception, understanding, and interpretation of the purported historical and constitutional compact between the state and the citizenry. This compact, briefly put, involves the citizenry's acceptance of a modicum of police power exercised in exchange for the orderly conduct of civil, legal, and spiritual life in both the public and private spheres.

The tentative conclusion to be drawn here is that, beyond a general consensus on the necessity of the sovereign powers of the state, there is little agreement on the particularities of that power and the domains of individual activity to which it should apply. This suggests that serious differences remain, especially over questions concerning the proper object of the rule of law, and particularly with regard to the civil liberties guaranteed to the population. These differences generally turn on the question of whether the rule of law concerns the primacy of the institutions of property, or whether the legislator or regulatory entity ought to aim for social justice, equality, and the general welfare as objects of law. This has led to increasingly pitched debates among citizenries globally, mostly along ideological lines, over whether the role of the state as it enforces the rule of law should or should not institute enabling mechanisms that would allow citizens to engage in market transactions without what one party to the debate perceives as the interference of the state.

Put in ideological terms, liberals and libertarians in the US take the rule of law to mean a "government of law and not of men." This view has greatly influenced the collective understanding of "good governance," which now generally connotes a democratic government that grants fundamental rights and liberties to its citizenry as its guiding principle (Wacks 2005). But these parties continue to disagree over the basic requirements of rights and liberties, particularly over the extent and purpose of restrictions on the regulatory and police powers of the state. In any case, formally speaking, laws and the rule of law are not authoritative simply because they are sanctioned by a sovereign. They are authoritative because: a) the legislator is of the citizenry and therefore under the same responsibilities and obligations as other citizens; b) the citizenry recognizes that no person suffers in body or goods except as a consequence of a breach of law; and c) the determination of the constitutionality of the law and breaches of it are made by properly constituted bodies.

This formal view of the rule of law has, of course, been amended and differentiated over time. From a liberal–utilitarian perspective, for instance, the value and legitimacy of law resides in its ends. Laws are thus legitimate because they promote the greater good of the greatest majority under the condition of equal liberties for all (Austin 1995; Bentham 1970). On this account, legal rules and processes must be based on facts and held to be valid and effective by their subjects. Laws must confer competence on individuals to pass judgment or enjoy given rights. They must establish clear criteria by which the validity of rules are recognized and accepted as appropriate to their purpose (Hart 1982; Kelsen 1967).

But, whereas libertarians and traditional liberals may view the law as an autonomous system of rules and principles, and so the rule of law as a neutral state of equity, legal realists have come to different conclusions which are relevant to the arguments of this chapter. Legal realists argue, in the first instance, that laws largely reflect political, social, and moral sensibilities. Laws therefore reinforce unequal power relations, disparate and contradictory interests between the governors and the governed, and therefore only embody the values of the former party. This fact can be observed, it is argued, in the predilections of judges and other adjudicatory officers.

Indeed, legal theorists commonly concur with these realist assumptions and hold that power and the economy play significant roles in both legislation and adjudication; that judicial outcomes depend on the connections and powers of persuasion of lawyers as well as on the sensibilities of judges and their affinities with supplicants; that private interests and the common good do not necessarily coincide; and that the attainment of the common good requires legislators, judges, and others to actively aim to supplement legislative and legal aporias for durable results toward an equitable legal system and, ultimately, a stable social order (Duff and Green 2005). Communitarian theorists and legal sociologists have nevertheless amended these views. John Rawls (2005) famously claimed that positivism and utilitarianism condoned social inequalities. According to him, the purpose of the rule of law was to advance social justice. Rawls forwarded two key principles of democratic entitlements as the foundations of the rule of law. The first was the promotion of equal rights to basic liberties in a context compatible with a system of liberty for all. The second Rawlsian principle was the fair and equal opportunity provided to all with the greatest benefit to be given to the least advantaged. Rawls's principles are also compatible with the views of legal sociologists. They, like Rawls, were preoccupied by social justice and thus looked beyond the positivity and utility of law. Following Durkheim (1964) and Weber (1978), legal sociologists have focused on the social function of law as either a source of legitimacy, a producer of social solidarity, or both. Accordingly, laws are systems of social rules and standards that create the normative contexts for individual entitlements and obligations toward others.

These interpretations of the rule of law came into play in the 1950s and '60s, when the concept appeared as "knowledge practices" under various schemes of modernization and Western legal assistance to developing countries. Initially, modernization and its related theories and practices of the rule of law favored strong, proactive states, vested in the development of a society and distributive justice within it. Western states and national development agencies reconciled public value and private ends through ideologies of duty and citizenship that established an equilibrium between the welfare state and private liberties as freedom. This equilibrium also favored a regulated economy. The resulting welfare state and associated modes of governance drew on normative links between popular sovereignty or self-determination and democracy, accountability, and entitlements (Trubek and Santos 2006).

In practice, however, formal laws did not necessarily imply the justness or legitimacy of laws themselves. For instance, in communist countries, constitutional prohibitions against arbitrary rulings did not lead to greater transparency. Likewise, modernizing authoritarian regimes in the developing world instituted rule-of-law environments that did not produce public accountability or individual liberties. The related legal systems were contrary to liberal democratic forms of government and the human rights they claimed to guarantee. This global state of affairs formed the basis of academic criticisms of the ideologies and practices of the rule of law in the 1970s and '80s. Then, as now, critical legal theorists, anthropologists, sociologists, and others noted that the implementation of formal legal systems alone had not ensured economic justice and good governance. Correspondingly, formal legal systems ceased to be the primary basis for understanding the principles of the social compact. Indeed, critical theorists objected that legal assistance and the theories behind it had been founded upon a simplistic model of history (Hatchard and Perry-Kessaris 2003; Trubek and Santos 2006). They variously showed that the abstracted model of rule of law was in need of rectification to account for spatially defined notions of values held by self-defined or self-determining collectives or nations (Huscroft, Miller and Webber 1988).

By the late 1980s, the combination of criticisms of rule of law programs and poor outcomes under them led to the loss of faith in liberal legalism and legal assistance. But this retreat was short-lived. The end of the Cold War, the advent of globalization, and the ascent of neoliberal institution-building brought to the fore market forces and market-friendly administrations in the US and United Kingdom that sought to rid the development process of politics and moral considerations, including distributive justice (Strange 1996). No longer constrained by alternative ideological models, the Reagan and Thatcher administrations associated market regulations with abuses of government power, and identified society and government as the causes of the corruption of public life and private morality. This new attitude toward government, as is well known, flew directly off the pages of F.A. Hayek's *The Road to Serfdom* (1944) and Milton Friedman's *Capitalism and Freedom* (1962). Both authors offered indictments of the repressive policies and outcomes that they attributed to twentieth-century collectivism, communalism, and socialism or, in any case, government intervention in market relations to shape outcomes toward social ends. In place of these projects, these authors and the governments that implemented their principles offered robust defenses of capital, property, and the associated mechanisms and relations of production, distribution, and consumption, which they defined through the rubric of market efficiency.

What was at stake in these writings was neither the concept of the rule of law nor its desirability in public life and market relations. It was instead the status, orientation, and ends or teleology of the law itself and the rule(s) it implemented. To these neoliberals, the real problem was that the state position under the rule of law had gone far afield from what was to be its proper role. Freedom could only be restored if the role of the state was redefined such that the constitutional role of the state—and therefore its regulatory and coercive powers—were oriented toward the enforcement of market rules (for instance, those applicable to contractual relations), the protection of individual assets (including property) and the preservation of public order. These kinds of constitutional forms were the only proper context and condition of freedom. Without any sense of irony, neoliberals had no aversion to greater regulatory and police powers. These forces simply needed to be properly deployed to keep workers, consumers, and non-market entities in check, to make them endure any injuries and disenfranchisements flowing from market relations and mechanisms as a matter of moral necessity. A "smaller state" came to mean targeted institutional atrophies of state capacity, to be augmented through the increase of police forces, prisons, and the military. The unleashing of police and military powers in this sense was indispensable to maintaining public order and therefore morally acceptable.

This complex of institutional arrangements and ideologies characterizes neoliberalism. Neoliberalism, thus understood, is liberalism turned upside down, with a reverse teleology in which the common good is defined away from public ends toward private interests as a matter of justice. This underlying ideology was adopted by states throughout the world through either perceived necessities or through mandates which were imposed by organizations associated with global capitalism (for instance, as "structural adjustments"), beginning with the World Bank Group (WBG) and International Monetary Fund (IMF). One need not believe in a "Washington Consensus" to note that market and capital-related bureaucracies from Washington, to Europe, to the UN, have come to favor the retreat of the state from the economic sphere in favor of corporations and the reduction of the role of civil society in economic decision-making. These bureaucracies and their technocrats now regularly propose to developing countries schemes of governance that skew political and economic power away from states and citizens toward capital, particularly global finance and corporations. This was the context in which the WBG entered into the institutional arrangements that led to the financing and exploitation of oil in Chad (Bantekas 2005). It is also the reason that one should be suspicious of the idea that the Chad Pipeline Project would lead to poverty reduction, equitable resource allocation, and capacity-building (World Bank 2006).

Today, critics of neoliberalism are quick to point out that the new ideology did not do away with the liberal model of history and its rationality, which primarily consisted in ridding the development process of political contestation (Hatchard and Perry-Kessaris 2003; Trubek and Santos 2006). Critics also note that legal assistance is currently oriented toward enforcement and normativity over ethical possibilities based on the contingency of legal events. Yet, neoliberal legal systems are contingent upon the particular industries that serve as their central nodes. For instance, Harvey (2003) has shown the tendency in global programming for extractive industries to undermine the legislative and regulatory powers of the state in favor of specific property and capital regimes. In the petroleum industry, the requisite legal regimes grant vast privileges and dispensations that create new private powers for the oil industry by transferring to it previously public legislative, regulatory, and administrative powers. Related institutional arrangements perversely augment the coercive powers of the state over the citizenry in order to enforce compliance with the instituted norms of expropriation and capital accumulation—whether in the form of contracts, force, oppression, or "looting." The wisdom or rationality of this sort of social engineering, that private agents are better suited to create the common good, is in question in the Chad experiment.

As the criticisms of it suggest, the neoliberal form of the rule of law enacted in Chad should be investigated before accepting its assurances that it offers guarantees against corruption, ensures poverty reduction, and, more importantly, preserves the interests and social and natural environments of local populations (Guyer 2002). While it may be credibly stipulated that legal regimes instituted as the necessary condition for the oil and pipeline project initiated the protection of property and investment rights in manners that are new, no sufficient relation between these regimes and good governance is evident. Early studies would seem to call into question any *prima facie* presumption that the interests of the WBG, the oil consortium, and the state coincide with those of the presumed beneficiaries of modernization (Amnesty International 2005; Rosenblum 2000). Rosenblum (2000), for instance, has illustrated legal deficiencies in the Revenue Management Plan that governs the distribution and uses of the oil revenues. Amnesty International (2005) contends that contractual clauses in the Environmental Management Plan (or EMP), which details methods of operating in conformance with environmental safeguards, give favorable treatment to the oil consortium that effectively undermine the regulatory authority of the state and the right of populations to obtain justice (33–36). Thus, I argue that the utility, efficacy, and legitimacy of the rule of law must be demonstrated, not simply assumed.

These studies suggested that the diction, or the wording, of the law, as well as the discourses around the benefits of law and the principle of legality did not seem to coincide with what they anticipated to be the effects of the instituted legal systems. Some of these effects became evident only after the judicial interpretations and bureaucratic applications of the law. According to Kennedy (2003), Trubek (2006), and Wacks (2005), for instance, the judiciary and other adjudicating offices are often swayed by litigants and applicants, or actors and agents that are able and willing to orient the operations of rules and procedures toward the protection of parochial interests. Speaking directly to the faith erroneously placed in the new legal experiments under neoliberalism, Kapur (2001) adds that the subjects of the new legal systems were often no more inherently dependable forces for justice than the oppressive states for which they were substituted. Likewise, they were no more responsive or accountable to the citizenry than the state and other state-instituted organs of adjudication. Nor was it self-evident that global actors such as the WBG or corporations produced better forms of governance and transparency than the institutions for which they were substituted. In sum, law, legality, and the rule of law must be considered not merely in light of the declared intentions of their initiators. Indeed, law, legality, and the rule of law are made of pretenses, predicates, and assumptions whose internal structures, or rules of constitution, particularly in regard to their assignments of privileges and immunities to legal subjects, are better predictors of the forms of justice that flow from them. This has been especially true in Chad.

This is why Strange (1996), Harvey (2003), and Campbell (2005) are justified in arguing that institutional schemes necessarily invite questions about the capacity to meet their declared ends. This means that the capacity of the regimes of law implemented in Chad to deliver the promised social or public goods of poverty reduction and more predictable juridical grounds for social interactions and activities were not simply a matter of diction or intention alone. In any context, the rule of law necessarily actualizes specific rules and procedures commensurate with explicit or assumed legal subjectivities and prerogatives that effect a range of legal rights, judicial endowments, and moral entitlements (Held and Koenig-Archibugi 2005). This is the case whether the circumstances of the enactment of the rule of law and incipient legal regimes constitute a transitional phase to modernization and economic development or not. It nevertheless remains the case that matters are more complicated and the risk for injustice magnified during periods of transition from one form of rule of law to another, especially when the new form negates any merit or validity to the previous one—as was the case in Chad. In such a case, according to Bromely (1986), Gasper (2004), Duchrow and Hinkelammert (2004), and Verdery and Humphrey (2004), one should anticipate the consequences of wholesale transfers of exogenous legal rules and processes to any new environments to be substantively considerable and therefore in need of examination. Such gestures as wholesale transfers of regimes of law are necessarily perceptible, if not measurable, because institutional innovations require shifts in the formal structures of governance. They also may insert novel ethical codes and socio-cultural rules of conduct or behavior that necessarily transform social relations and, in so doing, affect conceptions and the distribution of social goods.

## The nakedness of the rule of law

It is a central hypothesis of this chapter that the conventional associations between law and rule, on the one hand, and the rule of law with good governance, on the other, are mistaken in an era when finance capital gains ascendency over the state and in the process seeks to undermine its regulatory capacity. Until recently, the rule of law was associated with liberal democratic procedures as indispensable to progress (Tamanaha 2005). This view of the rule of law as an instrument of good governance appears in Latin America and elsewhere during constitutional transitions from authoritarianism to democracy. It was held that the rule of law ensured equitable and durable development in the interest of all. This outcome was inevitable, it was understood, because the rule of law eliminated corrupting influences in public life, preserved liberty and property, and thus created a propitious environment for foreign direct investment (FDI) (Chavez 2014).

This sense that the rule of law led to good governance spread rapidly with the successes of well-regulated authoritarian regimes, particularly among the so-called Asian Tigers. This moment produced the faith that good governance, if not democracy, would deliver progress. This was the moment of the first serious dent in the liberal orthodoxies of the connection between democratic and good governance and democracy and progress. Lee Kuan Yew famously claimed that the success of this "Asian model" was predicated upon "Asian values" (Sen 1997): ones that favored equitable development over democratic competition. It is now a matter of record that the concept of "Asian values" was conjured up and was more appropriately an ideological veneer than an ethnographic truth that was used to disguise the political undersides of the supposed Asian miracle. In retrospect, that miracle only gave temporary respite to authoritarianism as democratic forces would later compel governments to open up to political competition and respect for human rights (Riegel 2000). In Latin America, it was Chile that came the closest to the Asian model of law and order. This time, however, the road to the free market was murderous repression—as was the case in Indonesia—against forces that had supported socialist experiments. Traditional liberals witnessed events in Chile with alarm. This was not the case for others. Chile's seeming success in creating a vibrant market economy propelled the so-called Chicago Boys and their mentor, Milton Friedman, to canonical heights among the proponents of an unrestricted free market.

By the 1990s, the rule of law had re-appeared as both a panacea and a palliative to the fiscal crisis of the welfare state (Carothers 1998; Hatchard and Perry-Kessaris 2003; Tamanaha 2005). In their advocacy of the free market, Reagan, Thatcher, and others successfully instrumentalized the collapse of the Soviet Union and the demise of the post-colonial developmental model to formulate a new end to the rule of law. The rule of law appeared in the US and the UK in this context as a diction that signaled its own telos. This telos was often obscured by stylized representations that were intended to persuade wary audiences disenchanted with the "burdens" of "excess" taxation and "poor" public services. In truth, the diction of the rule of law was enunciated in two separate theaters with distinctly different effects. The end of its electoral enunciation was to weaken prior electoral coalitions in the West that favored the welfare state. This enunciation took the form of a cultural war against minorities ("welfare queens" being a classic neoliberal bogeyman) and the poor generally, both of whom presumably languished in a culture of dependence. In the realm of policy, however, the rule of law was intended to simultaneously dismantle regulations that impeded good market performance and, in their place, enact new regulations that protected corporations against the encroachment of the state and its citizens. The rule of law thus operated like a syntax that structurally connected seemingly isolated and disjointed laws, regulations, and rules into coherent and effective instruments on behalf of capital. Neoliberalism functioned in this context as a grammar that provided the structure that motivated legislators and oriented their actions toward their aim: the coming into being of regimes of edicts that generate legal privileges for investors, immunities for corporate operators, and legal protections required to secure property and prevent "frivolous" lawsuits by citizens.

One does not understand the rule of law, therefore, by studying individual pieces of legislation, administrative edicts, or judicial injunction. To know the rule of law is to know the object or motive of the assemblage of laws, edicts, and injunctions that apply to a particular object in any particular context. It is to investigate the relations between laws and regulations and the moral predicates underlying every mechanism, instrument, and process introduced as function of the rule of law. It is, in short, to demystify the internality of every actionable diction or injunction and their implied forms of subjectivity or, in other words, the manner in which subjects of law are positioned vis-a-vis legal objects. This subjectivity is the structure of legality (or not) that affects the actions, the nature, and the forms of adjudication to which legally recognized and sanctioned entities are subjected. In this way, the study of the constitutive dimensions of the rule of law must necessarily take into account the manners in which legislation is formulated and comes into form as law. It also therefore includes an examination of the stipulated intention of the law with respect to its consistency with the effects it produces. Finally, one must pay attention to the rules, procedures, and jurisprudence governing access to the instances of adjudication as well as those applicable to adjudication, including judicial processes.

The rule of law appears to provide a transformational grammar that opens up language (its enunciations) to a finite or limited number of possibilities (Chomsky 1957). This grammar forecloses the possibility of alternative forms of legality, or systems of legality, based on foundations that are not commensurate with itself. This is why the diction of the rule of law must be demystified by stripping it from tone, formulation, or style of delivery. It must be scrutinized on the basis of its underlying order or system of harmonious alignment between the whole of its internal rationality and the particular interests, desires, and values by which that rationality is driven. This is the only way to give sense or meaning to the rules and procedures of the law and to the necessity of each and every enchained string that is part of the rule of law. This required task is the separation of dogma from *episteme* such that the dimensions of the rule of law that obviously enchant us, such as individual liberties, are contrasted with instances of disenchantment: that moment when the individual is confronted with legal problems and corporate injustices with mechanisms or instruments of redress. These mechanisms or instruments are the determinants of possible outcomes in that they reveal the facilities accorded to the protagonists and antagonists of the law as well as the jurisdictions and legal bases of adjudication of states, their courts, and local institutions of justice.

## Mood, tone, and atmosphere

It is in the context of such enchantment with the rule of law that juridical interference with the existing operations of law in Chad was presented as legal assistance and salutary. Simply put, the tone, or general character and attitude evident in the discussions, was calibrated to convey a number of impressions: care, attentiveness to concerns, and determination to do good. This tone was adopted to convey to international and national NGOs and other concerned entities who were suspicious of the likely outcome of the project that the WBG would provide the assistance needed by Chad. The WBG also promised that it would ensure a reasonable allocation of oil revenues by providing a template for a Revenue Management Board (RMB) of government officials and members of civil society; that this board would see to it that oil revenues would be allocated according to the agreed priorities of poverty reduction, education, and infrastructure; that 5 percent of oil revenues would go to development projects in the oil-producing region; and, finally, that an independent International Advisory Group (IAG) would monitor progress along the way.

The pretense of the harmony of concerns and interests between the WBG, the state, the oil consortium, and elements of civil society made it almost ill-advised to ask questions of the nature of the proposed regimes of authority, property, and their modes of adjudication. Who would object to care for the poor? To the direct delivery of compensation to expropriated villagers? To an interceding third party (the IAG) to ensure the equitable dispensation of the rule of law? Indeed, the initial mood was one of unanimous relief. First, the good fortune of oil had occurred to an arid and poor country, decimated by years of conflict. Second, the World Bank had agreed to lend the necessary funds to the state to the get the project under way. Third, the World Bank had also agreed to maintain a presence throughout the duration of the project, and to actively play both an advisory and supervisory role. To match the mood, transnational environmental and human rights groups pinned their hopes on the promises of the World Bank, the oil consortium, and the government of Chad to allow for greater consultation and participation in the decision-making process. This process would also abide by international standards of human rights. Some local non-governmental groups, too, were initially elated by the promise of participation in different aspects of implementation of the oil and pipeline project.

The enthusiasm quickly dissipated as the terms of engagement among and between different actors emerged. First, the government introduced several decrees and edicts that specified the rules of circulation and assembly in the oil field region that divided security functions, with respect to jurisdiction and mandates, between the military police (*gendarmerie*) and the security guards for the consortium. Second, the state and the consortium agreed to a number of contractual dispositions that limited workers' rights and stipulated government liability for disrupting oil production. Third, the EMP was established as the final word for all non-technical matters regarding the exploration, exploitation, and transportation of oil. It contained, among other things, the terms of the expropriation of land, the manners of determination of ownership and compensability, and the categories of property subject to compensation. The EMP also introduced and defined the role of its local officers: including the *Agents Fonciers* (land agents) and Community Liaisons, to both disseminate information from the consortium to the population and to take the population's concerns to the consortium. Finally, the World Bank presented the list, jurisdiction, and procedures of the IAG.

These actions set the tone for the kind of interactions that would follow. They transformed initial optimism into wariness. In the oil region where the initial elation was subdued and reserved, the mood quickly soured by the tone set by different agents of the consortium—ExxonMobil, Petronas, and Chevron. Farmers were particularly concerned by the terms and conditions of expropriation, compensation, employment, and the securing of corporate assets. These were set by agents of the state, the corporations, and the World Bank with little consultation with populations or the organizations that represented them. The EMP, which summed up the commitments, responsibilities, and privileges and immunities of the different parties was therefore conceived from above, and so would be executed on contested terms. The first shock to the population was the realization that one of the corporations, through one of its experts, had managed to convince itself and others of one Chadian ontology that set town against nature (*wala*) with no fixed and determinate notion of property. This distinction entered into the moral calculus according to which compensation was allocated on the basis of the cultivation and/or utilization of the land for over 18 months.

These early revelations induced a sense of fear. The fear was justified, even if those who experienced it could not and did not predict the environment that would set in once the project was well underway. Once underway, many in the oil-producing region confronted the reality that the factors that decisively affected the nature or outcome of the rule of law were not in their favor. The core interests, values, and desires manifest in the EMP and state edicts clearly favored the oil companies, the state, and the international institutions. This was evident in the judicial and juridical facilities accorded to the World Bank, the state, the oil consortium, and the national and local offices to which they devolved some their powers and authorities. These facilities appeared in the privileges and immunities to define what was to be taken (land, or even the right to circulate in the oil region); what could be given away (jobs and discarded corporate junk); the conditions of expropriation (as indicated by instructions given to local chiefs, military police, county seat officials, etc.); the determination of the nature of the items and the price of the items to be compensated (including the values of different species of trees); the terms of cessation of employment; the values for compensation, whether for land or for labor; the conditions of the payment of compensation; and the nature of things that may be adjudicated as well as the forums for adjudication and conflict resolution.

We can deduce a number of observations from this. The first is that the source of law and the force of its enunciation are as important as the procedures and processes by which they take effect. Second, the similarities between or concurrences of law in time and space does not necessarily foretell universality. Nor does it imply universalism. This is to say that the particular edicts, injunctions and their derivations contained in the rule of law do not necessarily acquire universal value simply because they favor transnational corporations and financial institutions everywhere. If and when they are said to do so, it can only be in relation to the ascendency of the will and the desire of corporations and associated financiers to protect their investments. This is a different kind of application of the concept of universality than one that assumed a public good that is common, or one that is shared by investors, capital, civil society, and all elements of the larger society.

The rule of law therefore appears as social project that in modern societies assumes a certain number of formal and universal predicates behind which a secondary project necessarily lurks as its end. Under neoliberalism, this end is the survival and expansion of the market on behalf of capital. Poverty alleviation or reduction appears in this context formally, as the substantive universal good without which the expropriation of local populations or the spoliation of their environment could be justified. In this new climate, rather than producing development—whether equitable, sustainable, or something else—the ambition for and of the rule of law is more modest, as it was in the Chadian context. The rule of law was enunciated in this way to indicate that the pipeline project would: a) protect the interests of the poor through the institution of good public and corporate governance; b) institute effective social and environmental policies; and c) respect the human and customary rights of the populations served by the oil project (World Bank 2003). In actuality, the only attainable goal of the implemented rules and directives was to produce an economy through the law and related institutions, guaranteed by the exploration, exploitation, and transportation of oil according to terms that are favorable to the oil consortium.

Finally, the rule of law is an instrument of the mobilization of power and of extraction of resources. These resources are not merely material. They are first and foremost symbolic. In Chad, the rule of law was initiated at the behest of capital, disguised as a responsible solution to a number of international, national, and local supplicants on behalf of Chadians. In this set up, the WBG interceded on behalf of the International Finance Corporation (IFC) to ensure that a loan to Chad was a prudent investment. A member of the WBG, the IFC is self-described as "the largest global development institution focused exclusively on the private sector in developing countries" (IFC 2017). According to its webpage, in promoting investment in the private sector, the IFC nonetheless seeks "to help create opportunity for all" (IFC, 2017). It claims to join in this endeavor other institutions in the WBG whose products and services it leverages to "to provide development solutions customized to meet clients' needs" (IFC, 2017). In backing this claim, the IFC boasts not only its own financial resources but also "technical expertise, global experience, and innovative thinking to help our partners overcome financial, operational, and political challenges."

## Conclusion

The advent of neoliberalism signified a moment when finance capital gained ascendency over the state and, in the process, undermined the latter's regulatory capacity as well as restructured its social relations and sense of the common good, including entitlements, rights, justice, and equity. Even in the midst of the initial euphoria about the prospect of the Chad Pipeline Project, the outlines of the rule of law seemed ominous to the trained eye. It was clear, even then, that the WBG was interceding more on behalf of capital; that the sovereign will of the people was being sidelined through the marginalization of the state itself; that the principle of democratic governance and legislative authority of the people was disrupted; that the capacity of civil society to reproduce itself was being undermined; that the independence of the judiciary would be irrelevant to the activities of the consortium; and that the capacities of citizens to make claims for injuries arising from the activities of the oil consortium would be diminished.

The power of the ideological predicates of neoliberalism—freedom, property rights, and communal self-reliance—created the veil that initially masked these inevitable outcomes. The veil was effective because the material conditions stipulated as bases for World Bank intervention were real: Chad was a post-conflict state in need of a way to peace and stability through good govern-ance—and away from the corruptive influences of violence and economic mismanagement. This argument was sufficiently compelling to entice NGOs and boards of experts and advisors to play supplementary or subsidiary roles in the World Bank's efforts. The pretenses of sincerity and care by the World Bank helped to disguise the dismantlement of national and local institutions on behalf of capital, and to obfuscate the subordination of the interests of Chad and Chadians to those of capital.

This is why the apparent sincerity of the commitment to the rule of law and poverty reduction should not lead one to limit the analysis of the rule of law only to the extent to which its dispositions are carried out. The proper place to look is to the structures and implications of the regimes of edicts, dispensations, liabilities, and privileges and immunities that are articulated at the time of project design to define the contours of the possible and desirable. Indeed, laws are made of elements that have universal appeal in their enunciations, and yet each law is particular in its teleology and in its effects. This is to say that in Chad, the rule of was intended to preserve an illusion: that corporate capitalism, under market conditions, could more effectively deliver the welfare or development that the state could not. In actuality, it could not deliver on its assurances, because the law does not provide such an insurance. Instead, the rule of law has perverted the last vestiges of liberalism: democracy and citizenship.

We are therefore left with the replacement of the demos by a civil society that is not representative of society or the population. For better or for worse, civil society has analogically assumed the role that shareholders play in the market place: too diffuse and too disconnected to the real operations of the market to be useful—except on rare occasions. Presenting themselves as stakeholders, civil societies may in fact demonstrate an intensity of interests befitting any functionally specialized entity. Yet, this density of interests barely masks the intensity of the absence of influence by NGOs over their protagonists in the market and their internationally constituted forums. In the end, all they can do is hope for the best: that wealth would be so abundant as to trickle down and thus alleviate poverty; that a transfusion of market culture would occur in due time to effect the necessary behavioral adaptations needed for populations to prosper under the new rules; and that monetized relations would not irreparably disrupt social relations such that it leads to social decay. It is the best that observers can hope for, because the current rule of law offers no other point of reference to do otherwise.

## References

Amnesty International. 2005. "Contracting Out of Human Rights: The Chad–Cameroon Pipeline Project." London: Amnesty International. www.amnesty.org.uk/files/pol340122005en.pdf.

Austin, John. 1995. *The Province of Jurisprudence Determined*, edited by Wilfrid E. Rumble. Cambridge: Cambridge University Press.

Bantekas, Illias. 2005. "Corporate Social Responsibility in International Law." *Boston University Journal of International Law 22*: 309–347.

Bentham, Jeremy. 1970. *An Introduction to the Principles of Morals and Legislation*, edited by J. H. Burns and H.L.A. Hart. London: Athlone Press.

Berlin, Isaiah. 1969. *Four Essays on Liberty*. London: Oxford University Press.

Bowen, Howard R. 2013. *The Social Responsibilities of the Businessman*. Iowa City: University of Iowa Press.

Bromley, Daniel W., ed. 1986. *Natural Resource Economics: Policy Problems and Contemporary Analysis*. Hingham, MA: Kluwer Academic Publishers.

Campbell, David. 2005. "The Biopolitics of Security: Oil, Empire, and the Sports Utility Vehicle." *American Quarterly* 57 (3): 943–972. doi:10.1353/aq.2005.0041.

Carothers, Thomas. 1998. "The Rule of Law Revival." *Foreign Affairs* 77 (2): 95–106. doi:10.2307/20048791.

Chavez, Jenina Joy. 2014. "Why Nations Fail: The Origins of Power, Prosperity and Poverty." *Philippine Political Science Journal 35* (1): 124–127. doi:10.1080/01154451. 2014.907766.

Chomsky, Noam. 1957. *Syntactic Structures*. The Hague: Mouton.

Duchrow, Ulrich and Franz Hinkelammertranz. 2004. *Property for the People, Not for Profit*. London: Catholic Institute for International Relations.Duff, Antony, and Stuart P. Green, eds. 2005. *Defining Crimes: Essays on the Special Part of the Criminal Law*. Oxford: Oxford University Press, 2005.

Durkheim, Émile. 1964. *The Division of Labor in Society*, translated by George Simpson. London: Collier-Macmillan.

Friedman, Milton. 1962. *Capitalism and Freedom*. Chicago: University of Chicago Press.

Gasper, Des. 2004. *The Ethics of Development*. Edinburgh: The University of Edinburgh Press.

Guyer, Jane I. 2002. "Briefing: The Chad-Cameroon Petroleum and Pipeline Development Project." *African Affairs 101* (402): 109–115.

Hart, H.L.A. 1982. *Essays on Bentham: Studies in Jurisprudence and Political Theory*. Oxford: Clarendon Press.

Harvey, David. 2003. *The New Imperialism*. Oxford: Oxford University Press.

Hatchard, John, and Amanda Perry-Kessaris. 2003. *Law and Development: Facing Complexity in the 21st Century*. London: Cavendish, 2003.

Hayek, Friedrich. 1944. *The Road to Serfdom*. Chicago: University of Chicago Press.

Held, David and Mathias Koenig-Archibugi, eds. 2005. *Global Governance and Public Accountability*. New York: Wiley-Blackwell.

International Finance Corporation. 2017. www.ifc.org/wps/wcm/connect/corp_ext_ content/ifc_external_corporate_site/about+ifc_new.

Kapur, Devesh. 2001. "Expansive Agendas and Weak Instruments: Governance Related Conditionalities of the International Financial Institutions." *Journal of Policy Reform* 4 (3): 207–241.

Kapur, Jyotsna, and Keith B. Wagner, eds. 2011. *Neoliberalism and Global Cinema: Capital, Culture, and Marxist Critique*. New York: Routledge.

Kelsen, Hans. 1967. *Pure Theory of Law*. Berkeley: University of California Press.

Kennedy, David. 2003. "Laws and Developments." *In Law and Development: Facing Complexity in the 21st Century*, edited by Amanda Perry-Kessaris and John Hatchard, 17–26. London: Cavendish Publishing.

Locke, John. 1980. Second Treatise of Government. Indianapolis: Hackett Publishing Company.

Rawls, John. 2005. *A Theory of Justice*, original edition. Cambridge, MA: Belknap Press of Harvard University Press.

Raz, Joseph. 2009 (1979). *The Authority of Law: Essays on Law and Morality*. Oxford: Clarendon Press.

Riegel, Claus G. 2000. "Inventing Asian Traditions: The Controversy Between Lee Kwan Yew and Kim Dae Jung." *Development and Society 29* (1): 75–96.

Rosenblum, Peter. 2000. "Pipeline Politics in Chad." *Current History 99* (637): 195–199.

Sen, Amartya. 1997. Human Rights and Asian Values. Sixteenth Annual Morgenthau Memorial Lecture on Ethics and Foreign Policy. Carnegie Council on Ethics and International Affairs, May 25, 1997.

Strange, Susan. 1996. *The Retreat of the State: The Diffusion of Power in the World Economy.* New York: Cambridge University Press.

Tamanaha, Brian Z. 2005. "The Tension between Legal Instrumentalism and the Rule of Law." *Syracuse Journal of International Law and Commerce 33* (1): 1–24.

Trubek, David M., and Alvaro Santos, eds. 2006. *The New Law and Economic Development: A Critical Appraisal.* Cambridge: Cambridge University Press, 2006.

Verdery, Katherine and Caroline Humphrey, eds. 2004. *Property in Question: Value Transformation in the Global Economy.* Oxford: Berg.

Wacks, Raymond. 2005. *Understanding Jurisprudence: An Introduction to Legal Theory.* Oxford: Oxford University Press.

Weber, Max. 1978. *Economy and Society.* Berkeley: University of California Press.

World Bank. 2003. "Striking a Better Balance: Volume 1. The World Bank Group and Extractive Industries." Extractive industries review. Washington, DC: World Bank. https://openknowledge.worldbank.org/handle/10986/17705 License: CC BY 3.0 IGO.

World Bank Group. 2006. "Chad-Cameroon Petroleum Development and Pipeline Project: Overview." Report No. 36569-TD. December. Washington, DC: World Bank and International Finance Corporation. http://documents.worldbank.org/curated/en/821131468224690538/pdf/36569.pdf.

# Part II

# Contested imaginaries and claims to resources

# 4 "We own this oil"

## Artisanal refineries, extractive industries, and the politics of oil in Nigeria

*Omolade Adunbi*

In 2011, Nigeria announced an amnesty program targeted at the Niger Delta insurgents who, for many years, had crippled the oil industry. The amnesty program was designed to end in 2012, but it has since become a permanent feature of the Nigerian state (Adunbi 2015). While the amnesty program succeeded, to a large extent, in mitigating insurgency in the Niger Delta, new forms of contestation between the state and Niger Delta youth have emerged to challenge the state over its governance of the region's huge oil reserves. This chapter investigates how the state and young men in the Niger Delta are engaged in a process that is reshaping extractive governance in Nigeria. The chapter suggests that what is happening in the creeks of the Delta is changing community relationships and state and corporate extractive structures. The chapter is an attempt to rethink contests around oil extraction by looking critically at the emergent oil infrastructure constructed by youth as competing with state oil infrastructures in the production of crude oil.

In this chapter, I suggest that we move beyond seeing such oil infrastructure and the selling of crude and refined crude oil as the theft of state oil and that we redirect our focus to the processes through which such infrastructure is built. This emphasis allows scholars to study how technologies of crude refining reshape extractive governance in resource enclaves, and how artisanal refining is able to compete with state structures of extraction. In this chapter, I also seek to move beyond seeing the existence of local extractive processes and oil infrastructures as mere "survival strategies" (Ugor 2013, 271) adopted by youth in response to a criminal state (Bayart, Ellis and Hibou 1999; Ukiwo 2011). I argue instead that artisanal processes and structures of extraction reflect innovation and hybrid forms of knowledge, including knowledge of extractive technology that emerged in the 1930s with the imposition of high taxes on imported gin and liquor by the British colonial authorities. Technologies of ogogoro (local gin) production, made popular by youth in the 1930s, have become a useful tool in the production of crude oil by today's youth. The chapter argues further that participation in this particular mode of oil production is engendering relational forms that disrupt the state's claim to ownership of oil resources while at the same time creating opportunities for youth to establish structures that strengthen networks and cultivate support for their activities within communities rich in oil resources. I suggest that the Niger Delta represents an example of the hybridization of technologies of power that is illustrative of competing notions of extraction in an oil enclave. In the Bodo community in Ogoniland, local youth operate a refinery constructed with local materials using local technology that is not only representative of many of such refineries that litter the creeks of the Niger Delta but tells the story of how oil infrastructures have become a contested field between the state and local youth in the Niger Delta.

## Artisanal refineries, infrastructures and the oil creeks of the Delta

Recent scholarship on the governance of extraction in the post-colonial state focuses on the effects of resource extraction on power circulations both in and beyond the nation-state, the constitution of post-colonial imaginaries, and the interrogation of development or the production of modern subjects (Adunbi 2015; Askew 2002; Bayart 1998; Chalfin 2010a, b; Ferguson 2006; Hicks 2015; Mamdani 1996; Mbembe 2001; Mitchell 2002; 2009; Shever 2012; Watts 2004). While such scholarship remains valuable in informing our understanding of the politics of extraction in post-colonial states, I expand the scholarship on the post-colonial state by moving away from the emphasis on the effects of resource extraction to paying attention to the processes through which extraction happens and how these processes play into articulations of power and the production of competition between actors within the nation-state (Adunbi 2015; Ellis 2016; Leonard 2016).

Resource extraction, what Appel, Mason, and Watts (2015) call "a global production network with particular properties, actors, governance structures, ecologies, institutions and organizations" (18), creates a distribution and global trade network that is constituted as a distinct mode of knowledge production. This form of knowledge production is assumed to be restrictive to its participants i.e., oil corporations, and in Ferguson's (2006) telling, one that creates gaps between it and the countries or societies it purports to serve—gaps that are strengthened by the corporation's control and management of the market of extraction in cahoots with the postoclonial state. In describing the centrality of the market to extractive practices, Mitchell (2002) considers post-coloniality beyond its literal sense, writing that it should be seen as "forms of critical practice that address the significance of colonialism in the formation and practice of social theory" (7). In this situation, the market is central to the formation of the economy, which in turn is central to the construction of modern capitalism. While some scholars focus on how capital reconfigures power in a universalizing and modernizing mode, reshaping and reconstituting the global, other scholars challenge the idea of capitalist enterprises as the only form of knowledge that propels progress and modernization around the world.

Economies shaped by oil-capitalism, in particular, have illustrated the ways in which markets limit democratic practices. For example, the mining of coal in North America, Mitchell argues, helped build a democratic tradition because the workers were able to organize, embarking on strikes that could cripple local and national economies, particularly because coal is transported overland and its extraction requires huge numbers of workers (Mitchell 2009, 2013). In contrast, oil requires more technology and fewer workers and is an industry dominated by expatriates (Appel 2012; Leonard 2016) who have little or no interest in establishing democratic traditions in extractive enclaves. Central to this argument is how the physical nature of coal and its mining processes not only helped to shape democractic practices but also the establishment of a market system centered around commodity production.

Mitchell claims that the inability of oil to foster "forms of carbon-based political mobilization" explains "the undemocratic politics of oil" (2009, 422). While overtly avoiding the oil curse that has been dominant in much literature on oil and other resource extraction in many postcolonial states (e.g., Auty 2001; Humphreys, Sachs, and Stigliz 2007; Ross 1999; 2004; Schafer 1994; Shaxson 2007), Mitchell's argument ultimately acknowledges that democracy is not fundamentally the same in form and function everywhere, thus oil extractive processes can be central to practices of governance. The oil curse logic tends to neglect how oil interacts with people who live within these communities of extraction who are daily subjected to different modes of governance practices by corporations, insurgents, and NGOs (Adunbi 2015). It is this form of relationship between oil and democracy that Suzana Sawyer (2004) calls "subaltern oppositional movements [that] proliferate precisely through . . . transnational processes" of oil exploration, extraction, and commercialization (16). As much as the materiality of oil constrains democracy, it also creates possibilities for the inception of democracy and/or democratic participation, often in marginalized and overlooked spaces, such as those of the Niger Delta. Sawyer's approach requires a more comprehensive understanding of democracy. Democracy is not only a form and tactic of governance employed by nation-states, but also a lived experience characterized as "the managing and disciplining of movements of people, capital, and resources," structured by spatially and temporally-discrete ideals of egalitarianism (2004, 50).

In understanding these governance practices that shape extractive enclaves, I propose we also need a critical analysis of how oil extraction transforms the landscapes of postcolonial states from rural agrarian economies to transnational economies. This form of oil transformation, I argue, creates a centralized economic and political system that connects the nation-state to the international commodity market (Adunbi 2015; Apter 2005; Humphreys, Sachs and Stiglitz 2007; Leonard 2016; Vitalis 2007). Since oil extraction requires a vast and complex infrastructure that is largely provided by transnational capital and located in extractive enclaves, this transformation therefore generates capacity to create new modes of contestation around oil infrastructure.

This form of contestation can be seen in the transformation of populations, deprived of what they consider to be their oil and the benefits associated with it, taking it upon themselves to claim ownership in a variety of ways, such as building oil infrastructure in the creeks, to compete with state claims to oil. Today, there are over 1000 artisanal oil infrastructures in the creeks of the Niger Delta being managed by youth groups (Ugor 2013; Ukiwo 2011). The category "youth" is a very fluid one because someone who is over 50 can sometimes claim to belong to the category. In many communities of the Niger Delta, youthfulness is not determined by age but by the ability to organize and participate in activities that redefine extractive enclaves such as artisinal refining projects (Adunbi 2015).

While multinational oil corporations such as Shell, Chevron, and ExxonMobil invested millions of dollars in building pipelines, barges and flowstations for the exploitation and transportation of oil resources in ways that connect their business to the international market (Adunbi 2015; Mitchell 2009, 2011; Rogers 2014), youth operating oil infrastructures in the Delta use locally derived materials to build makeshift flowstations, and drums and tankers to transport crude oil to their customers. I suggest that we rethink the notion of oil infrastructure to encompass a particular practice where oil infrastructure sanctioned by the state and operated on its behalf by multinational oil corporations through a joint venture agreement with the federal government of Nigeria competes with a makeshift oil infrastructure constructed for the purpose of refining oil by youth in the creeks of the Niger Delta.

In this chapter, I use the term infrastructure to describe a system of organization that revolves around the exploitation, production, marketing, and institutionalization of oil as a local and international business commodity that is at the heart of the state and communities where the commodity is exploited. In conceptualizing infrastructure this way, I move beyond physical infrastructure to see oil infrastructure as a superstructure that facilitates the production, organization and management of this commodity and its benefits by various actors within and outside the nation-state. Infrastructure therefore has resonance with what Brenda Chalfin (2016) describes as "a focal point from which to track day-to-day practices, experiments, improvisations and replications, along with the tacit forms of consent and coercion that frame them" (19). Thus, oil is the fabric that can weld and disrupt. At the same time, it organizes and builds relationships that structure daily practices by actors who engage in managing, organizing and using oil.

As the lifeblood of a nation-state and the communities where it is exploited (Apter 2005; Ferguson 2006; Mitchell 2011; Rogers 2014; Watts 2004, 2012), oil shifts from being mere black crude into a valued commodity whose importance is shaped by its capacity to build relationships and to transform individual lives and the life of a nation-state. As I will show, oil's capacity to build relationships is demonstrated by the kind of relationships that youth develop with kin in their communities, as they manage oil infrastructures in the creeks. Oil's capacity to transform the life of a state from an agrarian and subsistence farming economy to that of resource extraction is entrenched in its ability to create wealth for the nation-state (see for example Adunbi 2015; Apter 2005; Watts 2004, 2012). Such transformations result in a situation where resource-rich communities, conscious of the wealth derivable from their environment and denied access to the wealth oil extraction generates for corporations and the state, now rely on relationships developed with other extractors, in this case youth who operate makeshift oil infrastructure, to derive benefits from oil. The relationships that the youth develop with members of the communities are often based on kinship ties. These ties transform the youth who manage makeshift oil infrastructure into community overseers of a new structure of governance that is crafted, managed and deployed within the creeks of the Niger Delta. Aided by knowledge of local technologies, for example, knowledge about how to construct boats and canoes, and with the cooperation of oil-bearing communities, youth who operate, manage and extract oil in the creeks are reshaping community spaces in ways that structure local governance. The participation and cooperation of people from the creeks of the Niger Delta communities in this complex network strengthens the power of the youth who operate local technologies of extraction. Developing strong relationships with members of the communities help strengthen the position of the youth as providers of benefits that had eluded them for many years because of state and corporate control. As an informant told me, "we know the youth who operate the refineries. For many years, the corporations would come and take our oil without giving back and the state will support them. When the youth came and told us they can help us enjoy some of our oil, the entire community rallied around and threw their support behind them."[1] Many informants corroborated this and when asked about their perception of what the youth are doing, many would respond thus,

> The youth are not thieves. They are not rogues. It is the state and the corporations that are thieves and rogues. Who cares about the state and corporation anyway? As far as we are concerned, the youth are our sons and they are doing the right thing for us.[2]

Making reference to the youth as "our sons" reifies the relation of kinship between the youth and community members who see what the youth do as a benefit to the entire community. By operating these technologies of extraction, the youth insert themselves at the heart of the communities such that relationships developed with members of the communities alter forms of governance to reshape lives and frame everyday practices, thoughts, and culture.

## Kpoo fire, bush refineries: crude oil and the creeks of extraction and production

Bodo, located in the Gokhana local government area of Rivers State is one of several communities that make up Ogoniland in the Niger Delta region of

Nigeria. Bodo is a few kilometers from Port Harcourt, the capital city of Rivers State. Bodo is made up of about 62,000 inhabitants living in 35 villages. In the past, fishing and farming were mainstays of community members and created employment for over 60 percent of the population, with trading, metal work, masonry, and carpentry accounting for the remainder. The community is also host to Shell Petroleum Development Company of Nigeria and oil wells are noticeable in parts of the community, particularly in the creeks. The activities of Shell have rendered many inhabitants of Ogoniland unemployed as a result of loss of livelihood, polluted waters and oil spills on farmlands. This is particularly true for people in Bodo, who engage in fishing and other agricultural enterprises.

Bodo community was also one of the centers of mobilization by the Movement for the Survival of the Ogoni People, or MOSOP. In the 1990s, the late Ken Saro Wiwa, who was killed by the military regime of General Sani Abacha in 1995, led the organization in its battle against Shell and the Nigerian state (Okonta 2008; Okonta and Douglas 2001; Apter 2005). The Ogonis sought compensation from Shell and the Nigerian state over their polluted environment. In 2011, the United Nations Environment Program, UNEP, issued a comprehensive report on pollution in Ogoniland which suggested that "oil pollution in many intercidal creeks has left mangroves denuded of leaves and stems, leaving roots coated in a bitumen-like substance sometimes 1 cm or more thick" (UNEP 2011, 3). Today, in the midst of these polluted environments where fishing and farming have become extremely difficult, many youth have resorted to constructing and managing oil infrastructure with support from community members.

In the summer of 2015, I visited the creeks of Bodo where there are thriving artisanal refineries. The creek, located in the Bodo West oilfield where Shell operates, is a sleepy village with only one access road to the community. As I approached the village, among the first noticeable features of the village were posters announcing the deaths and funeral arrangements of community members. As one informant pointed out, "this is what we now see in this village every weekend, many of our youth and elders are dying as a result of the polluted environment."[3] The informant further justified his claim by referring me to the recently published UNEP report mentioned above. There were also signposts placed by the Nigerian National Petroleum Corporation (NNPC), the official state oil company, warning of the dangers of drinking water from boreholes. While these signposts were prominent in the community, informants told me that the state does not provide alternative drinking water, so they still rely on the boreholes despite the health hazards associated with them.

As we approached the main entrance to the town, two men, who were later introduced to me as members of an artisanal refining group in the community, welcomed my research assistant and me. They later led us on their motorbikes to where we were to board a canoe to the creek. One of the men, Naaton,[4] who said he was in his early thirties, had scars on his face and arm.

The other man, Biebari, had an AK47 tucked beneath his flowing robe. The men brought us to the bank of the Bodo River where we met others, including Kenule, who described himself as a veteran of the 1990s Ogoni struggle against Shell Petroleum, who were waiting for us to arrive. Anchored across from where we boarded the canoe was a bigger canoe nicknamed the "Cotonou canoe." Cotonou is the commercial capital of the neighboring Republic of Benin and the canoe is so named because of its use in transporting crude across the ocean to Benin for trade in the international market—a point I return to below.

As we approached the creek of Bodo West, the smell of crude oil was striking. There were makeshift tankers, drums, cooling spots, pipelines, and a fireplace for refining oil—all locally constructed. In Bodo, artisanal refining is called "kpoofire." As we approached one of the refining tanks, Naaton announced, "welcome to our refinery where we refine our crude oil for sale." Welcoming me to "our" refinery, Naaton reiterated the notion that crude oil does not belong to the Nigerian state but to him and his community. As I sat down with him and other youth for an interview, he described to me how the refinery functions and the mode of governance used to manage the oil infrastructure. As Naaton was describing the process, Kenule nodded in agreement while also occasionally reiterating some of the points made by Naaton. Three layers of governance can be deciphered in the management of the oil infrastructure in Bodo creek. These three layers are intertwined and interconnected.

My use of governance is an attempt at problematizing everyday practices that structure relationships and organizations anchored in economic and political choices at state, community, and corporate levels in ways that shape the environment and all resources associated with nature (Adunbi 2015; Bonnafous-Boucher 2005; Chalfin 2016; Ferguson 2005, Foucalt 1991). By problematizing everyday practices, I show some of the shape and character of managing oil resources and the political choices made in engaging in the practice of resource governance in ways that create alternative systems of production within extractive communities. This alternative system of production results in the molding, construction and reproduction of practices that confront and consolidate knowledge and power in a variety of ways that include building community relationships as a way to access resources, using local knowledge in building infrastructure that is aimed at accessing natural resources while at the same time using the resources for one's benefits and the benefit of the community, and contesting ownership of resources with the state and corporations. Thus, this form of contestation produces three layers of governance.

The first is associated with those who have stakes in the local oil infrastructure but no presence in the creek. These are people at the top echelon of the artisanal refining business who are sometimes unknown to the youth who carry out the day-to-day administration of the creek. The operational principle of this category resonates with the principle of indirect rule where the leaders can be heard but not seen (Adunbi 2015; Lugard 1907; Mamdani 1996). I call this category the near-invisible governance model. By the near-invisible model, I problematize the concept of invisible governance by suggesting that they might not be visible but that their actions are visible to those who live by their rules.

The second layer is the direct governance of the creek. This mode of governance is organized around the youth who operate the artisanal refinery business on a day-to-day basis. It is this second category that is at the heart of this chapter. These youth construct, manage, and operate the local oil infrastructure in Bodo and other creeks of the Niger Delta. The layer of authority in this second category is well structured. For example, Naaton, Biebari, and others perform surveillance roles, which enable them to take up the role of chief security officers of the creek. They both had to welcome me and every visitor sanctioned by their leaders to the creeks at the entrance to the community. There is also the leader of the creek who is sometimes called the "chairman" and sometimes the "god-father" and he is accorded a lot of respect. The appropriation of such titles as "chairman" and "god-father" illustrates the hierarchical system of operation within the creeks where the "chairman" is seen as the symbol of authority. The chairman not only presides over meetings but also, in consultation with his other cabinet members, makes business and administrative decisions for the group. To make the administrative duties of the chairman easy, he has a cabinet with positions such as vice-chairman, General Secretary, Finance Secretary, and Security Secretary.

The final layer consists of the community members who cooperate and support the two previous categories—the invisible layer and the direct governance layer. The cooperation and support of many community members is anchored in the direct benefits they derive from participation in the management of extraction in the creeks. Members of Bodo community participate in the governance of extraction in two ways. First, many serve as local distributors and marketers of refined petroleum products (Naanen and Tolani 2014; Social Action 2014; Ugor 2013). During my visit to Bodo, jerry cans filled with refined oil for sale were displayed in front of houses within the community. Some members would also transport their products to nearby cities for sale. By participating in this way, members of the community make the claim that they are benefiting from oil they "own." This also arguably improves the economic conditions of many community members who have been displaced from their livelihood by oil prospecting in the community.

The second mode of participation is based on the free kerosene that is distributed for domestic use. This is a form of what Rogers (2014) calls "petrobarter," which he describes as "discourses and practices that feature the exchange of oil for all manner of goods and services without the direct intervention of money" (131). Since many people have young family members who participate in the extraction process, they all become beneficiaries of this free kerosene, a by-product of refined crude oil used mainly for cooking. The price per gallon of kerosene is higher than that of petrol, so its distribution is an important way for youth to firm up support for their operations in the creeks.

Household members not only use free kerosene for cooking, sometimes they sell it in nearby communities as a way of earning cash to support their families. Free kerosene helps build relationships with community members in ways that guarantee loyalty, goodwill and support. In return for the kerosene that is freely distributed, many members of the community provide different kinds of support such as not providing any information about the activities of the youth to the state. As one informant mentioned to me during our interaction,

> sometimes the state will send their agents here to ask us questions about who is behind the refineries. We will respond with a categorical "no." Why should we give our sons and brothers away when we know they are doing the right thing for the community and us? We will never and can never do that.[5]

Community members help local operators to conceal construction of artisanal refining infrastructure projects from the prying eyes of the state, since resource infrastructures are the properties of the Nigerian state by the petroleum Act of 1969 and other subsequent laws, such as the Land Use Act of 1973 (Adunbi 2015; Watts 2004). In addition, many agents of the state are also complicit in the concealment of the construction of the refining infrastructures because the operators often pay them off. While the youth had the necessary skills to build and manage these infrastructures, the near-invisible participants mentioned earlier are in most cases the financiers of the projects. The chairman and a few of his cabinet members who are also privy to the presence of the near-invisible participants render returns in terms of profit made from sales to those participants.[6] Thus, the three layers—those who have direct control of the creeks, those who are not directly present in the creeks, and members of the community —all participate in a relationship that is defined by crude oil and entwined by benefits of its extraction.

## Artisanal refineries, oil infrastructure and the technologies of extraction and production

On my return from the creeks in the summer of 2015, I met with a group of United States Agency for International Development (USAID) contractors in the city of Port Harcourt. While I was sharing drinks with Niger Delta activists at a bar in my hotel, three men dressed in well-cut suits approached our table and asked if they could join us. The three men knew the activists because the activists had participated in the training programs organized by the contractors in the Delta. The training programs, I was told, were aimed at "empowering" youth to "think outside the box" and to embrace programs that could make them self-sufficient without having to rely on a supposedly "dysfunctional" Nigerian state that could not deliver services to its citizens.

The contractors were Nigerians but believed that the state should be pushed out of the way for innovation to thrive—the typical neoliberal sentiment that dominates development discourse from Washington. One of the contractors, who described himself as an "innovation leader,"[7] was the first to comment on the state of the Niger Delta. He suggested,

> the youth of this region are very lazy; they lack innovative ideas and all they know is how to carry arms against the state and corporations doing business in the region. If they were innovative, they would have realized there are more opportunities than oil but they all tend to focus on oil all the time.

This comment generated a vigorous debate. The activists argued that the contractors had limited knowledge of the crisis of environmental degradation and the absence of social amenities that, in their view, were needed to drive innovation in the Delta communities.

What aroused my interest in the conversation was the emphasis on innovation and the perception, shared by the contractors and vehemently opposed by the activists, that Niger Delta youth lacked the drive or the resources to innovate. In the last few decades, innovation has been a mantra of neoliberalism (Ferguson 2011; Ong 2006). Innovation resonates with what Ong sees as the attributes of neoliberalism, which is, "conceptualized not as a fixed set of attributes with predetermined outcomes, but as a logic of governing that migrates and is selectively taken up in diverse political contexts" (2006, 2). To the USAID contractors, innovation was the catalyst to growth and Niger Delta communities were not "growing" because they were not innovative enough to meet the predetermined meaning of innovation set by neoliberal standards. To the activists, there exists a clear distinction between innovation and growth. Many of the activists believed Niger Delta youth are innovative while the problem of growth lies at the doorstep of the state that had adopted wholesale neoliberal economic and political practices that stifle innovation and degrade the Niger Delta environment. While the debate was raging, I saw myself aligning with the Niger Delta activists by agreeing with their take on neoliberal economic and political practices which privilege corporations in their activities in the region (Adunbi 2015).

Therefore, contrary to the neoliberal dictum, what was missing in Niger Delta communities was not innovation, but drivers of innovation such as the state, which does not invest in education in many of these Niger Delta communities. This is why it is unimaginable that contractors to a state-funded agency such as USAID would make the argument that the state constitutes a cog in the wheel of economic growth in the Niger Delta. However, as I show below, even in the absence of state investment in education, innovation is not lacking in the Niger Delta communities, as proponents of neoliberalism would want us to believe. The building of oil infrastructure using local technologies is an excellent example of how innovation drives survival in the oil fields of the Delta.

While the state and its business partners engage in practices that disentangle communities from the benefits of oil wealth, many youth in the Delta, particularly the Bodo communities, have devised ways of deriving benefits from oil fields using local technologies. These technologies include the construction of Cotonou boats and speedboats from wood, makeshift refineries made from corrugated iron sheet, and fabricated pans as well as pipes and drums. Local welders, carpenters, and masons are instrumental to the construction of the oil infrastructure. Materials are sourced from within the community and nearby cities such as Port Harcourt. Welders convert metal into pipes that are connected to tanks where oil is refined and build tanks using corrugated iron sheet and metal. Carpenters construct the boats that are used in transporting drums of crude oil to the refining site. Because of the size of Cotonou boats, which have the capacity to hold between 100 and 600 drums of crude oil, they cannot be used to transport drums of crude to the refining site; therefore they are stationed at a site where navigation is easier. When crude oil is refined, speedboats are used in transporting it in drums back to where the Cotonou boat is located and the Cotonou boats are used to transport the finished product to consumers.[8] Many of the oil infrastructures, particularly tanks and pipes, are located at the refining sites in the creeks. The design of the refinery is such that two big tanks are stationed close to each other, both tanks connected to pipes.

In Bodo creek, there are about four such refining sites within a short distance from each other. One of the tanks functions as the site where the crude oil is deposited while the other tank serves as a reservoir for the refined product. There are about four pipes connected to the tanks; each serves its own function. For example, one pipe processes the crude into gas, another processes crude into diesel, the third processes crude into kerosene, while the last pipe transfers the waste product into a water pond that serves as a cooling tank. The tanks are elevated above a huge fireplace that performs the role of a refining burner. This is where the notion of kpoofire originates as the term refers to the process of throwing crude oil into the tanks where refining takes place. When crude is emptied into the tank, it makes a sound, kpoo, hence the youth christened the refining process kpoofire.

When it is thrown into fire, the crude is allowed to burn for hours unattended until it becomes a finished product after which it will be allowed to cool down before it is packaged into drums for transport to the Cotonou boats. Refining takes place at night and during the process of refining, the creeks appear to be lit up from afar. This process is similar to when a corporation's extraction processes flare gas into the atmosphere. Hence, the youth say that their refining process is similar to those of the oil corporations because at night, artisanal refining processes can be seen from a distance just like the gas-flaring of the corporations.

The youth access crude oil through a practice known as "tapping." Tapping denotes the process of perforating existing oil pipelines to access crude oil. This is where the job of a welder becomes important to the extraction process.

Many of the welders who helped in building tanks and pipes for the refining process are previous contract staff or employees of corporations such as Shell, AGIP Oil, or TotalFinaElf. The skills acquired by local welders in helping corporations build and manage pipelines have become important tools used in tapping oil for the artisanal refining process.

Crude oil is tapped from the pipelines at night and speedboats are used to transport the tapped crude to the refining sites where it is refined into the finished product using kpoofire. As one informant told me,

> we usually start our operation around midnight. We know the terrain very well. We have lived all our lives here and some of us had worked for the corporations as welders before, so we know how to open and caulk pipes. The process is easy, we open the pipes with our tools, load the crude into drums and then caulk the pipe back when we are finished.[9]

Tapping is not limited to pipelines alone. As Naaton explained to me, the youth also "break into wellheads, installing their own pumps and using hoses to load oil onto barges" (Social Action 2014, 24; Ugor 2013). Such operations demonstrate the dexterity and resourcefulness of the youth and their ability, despite the protestations of the USAID consultants, to innovate.

Many in Bodo and other Niger Delta communities who participate in the practice of oil refining combine a form of hybrid knowledge of local practices with the knowledge acquired through learning from the corporations to produce a distinct method of extraction and production of oil. The practice can be described as the art of caulking and piping through a process assimilated from corporate practices of extraction and combined local technologies and knowledge in siphoning oil from pipelines and flowstations in the Niger Delta. What becomes clear is the fact that technologies of extraction, innovative as they are, also borrow from local technological knowledge that has been with Niger Delta communities for several decades. Extraction in the Delta did not start with oil and it is this old practice of extraction that has also been innovatively embedded in today's extractive practices. As I show in the next section, it is the reuse of old technologies of distilling local wine and liquor such as ogogoro that helps consolidate today's refining processes.

## Reinventing old technologies: ogogoro and the techniques of refining and producing crude oil

Ogogoro, the local gin brewed in Niger Delta communities, provides a window onto understanding the process of crude refining in Nigeria. The local brew, also known as kaikai in Port Harcourt, Sapele water in the Warri area of the Delta, and "Craze man in bottle" in other areas, became popular during the colonial era (Heap 2008). The history of local gin distilling is told in two parts. The first part, as narrated by my informants, and closely observed during my many years of research in the Niger Delta, started when local brewers specializing in brewing palm wine used their products during the celebration of harvests, funerals, marriage ceremonies, and other important community achievements. The second is told from the lens of colonial history dating back to the 1930s (Heap 2008) when the colonial authorities imposed taxes on imported spirits, which made them unaffordable to many consumers.

In the 1920s for example,

> due to general prosperity in that decade, gin imports increased dramatically. But in the depressed economic conditions of the 1930s, when farmers found their export produce selling for unprecedentedly low prices and imported liquor priced beyond their reach by high customs duties, ogogoro supplied an immediate want.
>
> (Heap 2008, 576–577)

With the economic depression of the 1930s, the colonial administration saw imported spirits as an opportunity to raise revenue for the colonies and their home government; hence the high tariff on imported products. Heap (2008) suggests that attempts at brewing local gins started with "Stocky James Iso, a 35 year old native of Calabar" (2008, 576), who learned the "illicit art of distilling moonshine" (576) during his sojourn in the United States. On his return to his native Calabar[10] in 1924, he began distilling gin. Shortly after his return, Iso sold the secrets of distilling to others interested in brewing spirits around Calabar.

However, Iso's secret recipe generated interest, and as the Nigerian Echo Newspaper reported in the 1930s, "illicit distillation sprouted up like mushrooms overnight in Calabar province and it has spread all over the country to such an extent that there is hardly any province that can plead absolute innocence of this evil" (Heap 2008, 577).

The brewing of ogogoro became a way to display the innovativeness and ingenuity of the local fabricators. Blacksmiths, palm wine tappers and farmers who cultivated sugar cane provided the necessary skills and materials for the brewing of the local gin. Blacksmiths would make pipes and connect them to drums; palm wine tappers would provide the wine from palm trees, while farmers would provide sugar cane. In making ogogoro, local brewers connect a pipe to a drum, which will in turn be connected to another drum holding water before passing it into the third drum where the distilled gin will settle after burning in the main drum for hours. As Heap suggests, "treating the distilled alcohol with burnt sugar engineered a spirit similar in bouquet and appearance to whisky, while adding tobacco juice made it brandy colored" (2008, 582). The production of ogogoro gin, started in the Niger Delta in the 1930s, continues until today. There is hardly any community that does not know or has not heard about ogogoro and in many communities it is considered an essential drink at festivals, funerals, marriages and other celebrations.

However, the techniques of extraction deployed in the region to sidestep what locals considered obnoxious colonial laws prohibiting the production of local gin have become useful tools for circumventing the state and multinational oil corporations' extractive practices in the Delta today. Technologies of ogogoro extraction have been modified to facilitate the techniques of oil extraction. As Kenule told me during an interview, many of the youth in the region have, at one time or the other, been involved in the processing and extraction of ogogoro, so the technology is not new to the area. According to Kenule, "after fishing and farming, ogogoro production is a popular mode of earning income in our community. Many see this as a family business, so you are trained from early on in the process of producing it." Many informants corroborated Kenule's assertion by reiterating that ogogoro production is more like a family business in many parts of the Niger Delta. Of course, such evidence abounds in many communities because many of the households I visited had a small-scale ogogoro production business that they relied on as an income supplement for the family in addition to fishing and farming. Moreover, the shift from ogogoro production to that of crude has also meant a shift in technologies. The new technologies are much more sophisticated than the technologies deployed during the colonial period. Ogogoro production was mainly done by families and mostly within the household. However, oil production today has taken on a life of its own and the creeks epitomize not only its sophistication but also a system that encompasses networks of participants, including youth, community members, international networks, and local patrons.

More importantly, a parallel can be drawn between ogogoro production in the colonial period and today's artisanal practices of oil production. Both are (or were) considered illicit. Just like the colonial administration considered ogogoro an illicit trade because of its impact on the colonial revenue base, the Nigerian state and its business partners consider the local production of oil as the theft of state property because of its capacity to make a dent in state revenues. As Van Schendell and Abraham (2005) remind us, illicit, that which is socially perceived as unacceptable, can also be historically changeable and contested. While the state sees the extraction of crude oil as an illicit act that must be punished, the youth who have built their livelihoods around it contest the illicitness of their actions by reverting to how their own hybrid knowledge that combines the complexity of tapping and caulking with the practice of ogogoro production helps in navigating the sophistication of extraction. As one informant said, "we use our local knowledge base to turn crude oil into refined oil. The state and corporations will say we are stealing oil, but to us, we are merely tapping from what the state has illegally taken from us."[11] The sophistication of the crude refining techniques of today emanate from the knowledge base of youth who worked as employees or subcontractors of the oil corporations and local knowledge of ogogoro production. These former employees, in alliance with other youth who are skilled in boat construction, combine to provide a space where crude oil is turned into refined oil.

## Conclusion

In January 2015, the Royal Dutch Shell Company agreed to an $84 million dollar settlement with the Bodo community in acknowledgment of the environmental damage sustained by the region as a result of two oil spills, which occurred in 2008 and 2009 (BBC News 2015). The story is noteworthy not only because it represents the first major settlement to be reached between a major multinational oil company and the local community ravaged by the consequences of oil refining, but also in terms of how it contributes to the persistent narrative emerging about oil and the Niger Delta. This prevailing narrative focuses on the "failed promise" of oil, whereby oil, rather than serving as the "life-blood" of the state and the community from which it is extracted, impinges negatively on these communities in a myriad of ways that include environmental degradation and the loss of livelihood. First is the inability of the region to benefit from this hidden wealth, as most of the capital moves beyond the local community to multinationals situated in more developed nations and to the elites of the state. Second, oil often creates environmental damage in these communities, which is also reflected in most of the news and press coverage about the region. The third narrative strand centers on the socio-cultural consequences arising from the oil infrastructure, where illegal refining and resentment by local populations are cast as the primary villains and as the cause of much of the crime and environmental harm in the region. It is worth noting that even as Shell accepted liability for the oil spills, company executives were quick to assign blame for the environmental damage to the illegal refining process.

The problems that emerge from these failed promises of oil deserve acknowl-edgment and are a crucial and critical area of concern for scholars and activists alike. These narratives highlight the undemocratic forms of governance that prevail within extractive enclaves, and have thus been accorded significant attention. Yet, within the shadows of this overarching narrative of oil-as-unde-mocratic, narratives that are far more complex can be unearthed. The impulse toward binary thinking, which casts the illegal refiners against the multinationals, fails to account for how relationships are molded in ways that produce hybridized knowledge and new forms of governance within local communities. It is this hybridized knowledge that has turned the creeks, known for their access to fishing and farming, into sites of alternative systems of production of crude in the creeks which manifest themselves through natural resources and also push back against neoliberal conceptions of poverty being the result of a lack of innovation and agency. As this chapter shows, the Niger Delta has a long history with innovation and with technologies of extraction and production. Technologies of extraction, in the historical annals of the Delta, have been epitomized by innovation that drives resistance to outsider influence and dictates. Ogogoro distilling technology helped in shaping new innovation and techniques of extraction when the national economy shifted to oil. It is this innovation that produces the artisanal refineries that litter the landscape of the Niger Delta. More importantly, close analysis of the artisanal refining process refutes any simplistic dismissal of these activities as unorganized or fueled solely by brigands and bandits. Instead, the politics of crude oil governance in Bodo community reveals a solid and complex structure, embedded in local relationships that allows for an integrated amassing of resources, both technological and cultural, coming together to create a vibrant socio-economic infrastructure.

## Notes

1　Interview conducted at Bodo, Ogoniland on July 10, 2015.
2　This is a paraphrasing of interactions I had with community members in Bodo on the afternoon of July 10, 2015.
3　The conversation with an informant took place as our vehicle made its way to the center of the community.
4　All names used in this chapter are pseudonyms in order to protect the identities of my interviewees.
5　Interview conducted in Bodo, Ogoniland on July 15, 2015.
6　During the period that I was there, I tried several times to know the exact profit that they make and how it is distributed but they never shared the details of this information with me.
7　The other contractors were fond of calling him an innovative youth leader because of his dexterity at developing what they considered to be innovative programs aimed at tackling social problems. However his innovative ideas were never disclosed in our conversations.
8　I use "consumers" to denote those who purchase the finished products from the youth and not necessarily those who are the end users of the product. Sometimes the consumers are the middlemen who purchase the products and sell them to their international partners who oftentimes wait for their consignments offshore, in the Atlantic Ocean.
9　Interview conducted at Bodo, Ogoniland on July 10, 2015.
10　Calabar is the administrative headquarters of Cross Rivers State. The state lost the last of its oil wells to Cameroon in a judgment delivered by the International Court of Justice in 2002 at The Hague. The judgment transferred ownership of the disputed Bakasi peninsula to Cameroon and Nigeria effected the transfer in August of 2008.
11　Interview conducted in Bodo, Ogoniland, July 21, 2015.

## References

Adunbi, Omolade. 2015. *Oil Wealth and Insurgency in Nigeria*. Bloomington: Indiana University Press.
Appel, Hannah. 2012. "Offshore Work: Oil, Modularity, and the How of Capitalism in Equatorial Guinea." *American Ethnologist* 39 (4): 692–709. doi:10.1111/j.1548-1425.2012.01389.x.
Appel, Hannah, Arthur Mason, and Michael Watts, eds. 2015. *Subterranean Estates: Life Worlds of Oil and Gas*. Ithaca: Cornell University Press.
Apter, Andrew H. 2005. *The Pan-African Nation: Oil and the Spectacle of Culture in Nigeria*. Chicago: University of Chicago Press.
Askew, Kelly Michelle. 2002. *Performing the Nation: Swahili Music and Cultural Politics in Tanzania*. Chicago: University of Chicago Press.
Auty, Richard M. 2001. "The Political Economy of Resource-Driven Growth." *European Economic Review* 45 (4–6): 839–846. doi:10.1016/s0014-2921(01)00126-x.

Bayart, Jean-Francois. 1993. *The State in Africa: The Politics of the Belly*. London: Longman.

Bayart Jean-François, Stephen Ellis, and Hibou Béatrice. 1999. *The Criminalization of the State in Africa*. London: International African Institute in association with J. Currey, Oxford.

BBC. "Shell Agrees $84m Deal over Niger Delta Oil Spill." *BBC News*. January 7, 2015. www.bbc.com/news/world-30699787.

Bonnafous-Boucher, Maria. 2005. "From Government to Governance." In Maria Bonnafous-Boucher and Yvon Pesqueux, *Stakeholder Theory: A European Perspective*, 1–23. London: Palgrave Macmillan.

Chalfin, Brenda. 2010a. *Neoliberal Frontiers: An Ethnography of Sovereignty in West Africa*. Chicago: University of Chicago Press.

Chalfin, Brenda. 2010b. "Recasting Maritime Governance in Ghana: The Neo-Developmental State and the Port of Tema." *Journal of Modern African Studies* 48 (4): 573–598. doi:10.1017/s0022278x10000546.

Chalfin, Brenda. 2016. "'Wastelandia': Infrastructure and the Commonwealth of Waste in Urban Ghana." *Ethnos: Journal of Anthropology*. Special Issue: Infrastructures: 1–24. doi: 10.1080/00141844.2015.1119174.

Ellis, Stephen. 2016. *This Present Darkness: A History of Nigerian Organized Crime*. New York: Oxford University Press.

Ferguson, James. 2006. *Global Shadows: Africa in the Neoliberal World Order*. Durham, NC: Duke University Press.

Ferguson, James. 2011. "Toward a Left Art of Government: From 'Foucauldian Critique' to Foucauldian Politics." *History of the Human Sciences* 24 (4): 61–68. doi:10.1177/0952695111413849.

Foucault, Michel. 1991. "Governmentality," translated by Rosi Braidotti and revised by Colin Gordon. In *The Foucault Effect: Studies in Governmentality*, edited by Graham Burchell, Colin Gordon, and Peter Miller, 87–104. Chicago: University of Chicago Press.

Heap, Simon. 2008. "Those That are Cooking the Gins: The Business of Ogogoro in Nigeria during the 1930s." *Contemporary Drug Problems* 35 (4): 573–610.

Hicks, Celeste. 2015. *Africa's New Oil: Power, Pipelines and Future Fortunes*. London: Zed Books.

Humphreys, Macartan, Jeffrey D. Sachs, and Joseph F. Stigliz, eds. 2007. *Escaping the Resource Curse*. New York: Columbia University Press.

Humphreys, Macartan, Jeffrey Sachs, and Joseph E. Stiglitz, eds. 2007. *Escaping the Resource Curse*. New York: Columbia University Press.

Leonard, Lori. 2016. *Life in the time of Oil: A Pipeline and Poverty in Chad*. Bloomington: Indiana University Press.

Lugard, Frederick John Dealtry. 1907. *Northern Nigeria (Report for the Period from 1st January, 1906, to 31st March, 1907, by the High Commissioner of Northern Nigeria)*. London: HMSO.

Mamdani, Mahmood. 1996. *Citizen and Subject: Contemporary Africa and the Legacy of Late Colonialism*. Princeton: Princeton University Press.

Mbembe, Achille. 2001. *On the Postcolony*. Berkeley: University of California Press.

Mitchell, Timothy. 2002. *Rule of Experts: Egypt, Techno-Politics, Modernity*. Berkeley: University of California Press.

Mitchell, Timothy. 2009. "Carbon Democracy." *Economy and Society* 38 (3): 399–432. doi:10.1080/03085140903020598.

Mitchell, Timothy. 2011. *Carbon Democracy: Political Power in the Age of Oil.* London: Verso Books.

Naanen, Ben, and Patrick Tolani. 2014. *Private Gain, Public Disaster: Social Context of Illegal Oil Bunkering and Artisanal Refining in the Niger Delta.* (Technical Report). Port Harcourt, Nigeria: Niger Delta Environment and Relief Foundation (NIDEREF) and University of Port Harcourt. www.nideref.org/.

Okonta, Ike. 2008. *When Citizens Revolt: Nigerian Elites, Big Oil, and the Ogoni Struggle for Self-Determination.* Trenton: Africa World Press.

Okonta, Ike, and Oronto Douglas. 2001. *Where Vultures Feast: Shell, Human Rights, and Oil in the Niger Delta.* San Francisco: Sierra Club Books.

Ong, Aihwa. 2006. *Neoliberalism as Exception: Mutations in Citizenship and Sovereignty.* Durham, NC: Duke University Press.

Rogers, Douglas. 2014. "Petrobarter: Oil, Inequality, and the Political Imagination in and after the Cold War." *Current Anthropology* 55 (2): 313–353.

Ross, Michael L. 1999. "The Political Economy of the Resource Curse." *World Politics* 51 (2): 297–322. doi:10.1017/s0043887100008200.

Sawyer, Suzana. 2004. *Crude Chronicles: Indigenous Politics, Multinational Oil, and Neoliberalism in Ecuador.* Durham, NC: Duke University Press.

Shafer, Michael D. 1994. *Winners and Losers: How Sectors Shape the Developmental Prospects of States.* Ithaca: Cornell University Press.

Shaxson, Nicholas. 2007. "Oil, Corruption and the Resource Curse." *International Affairs* 83 (6): 1123–1140.

Shever, Elana. 2012. *Resources for Reform: Oil and Neoliberalism in Argentina.* Stanford: Stanford University Press.

Social Action. 2014. *Crude Business: Oil Theft, Communities and Poverty in Nigeria.* Port Harcourt: Social Action.

Ugor, Paul U. 2013. "Survival Strategies and Citizenship Claims: Youth and the Underground Oil Economy in Post-Amnesty Niger Delta." *Africa 83* (2): 270–292. doi:10.1017/s0001972013000041.

Ukiwo, Ukoha. 2011. "The Nigerian State, Oil and the Niger Delta Crisis." Essay. In *Oil and Insurgency in the Niger Delta: Managing the Complex Politics of Petroviolence,* edited by Cyril I. Obi and Siri Aas Rustad, 17–27. London: Zed Books.

United Nations Environment Program (UNEP). 2011. *Environmental Assessment of Ogoniland.* Nairobi: UNEP.

Van Schendel, Willem, and Itty Abraham, eds. 2005. *Illicit Flows and Criminal Things: States, Borders, and the Other Side of Globalization.* Bloomington: Indiana University Press.

Vitalis, Robert. 2007. *America's Kingdom: Mythmaking on the Saudi Oil Frontier.* Stanford: Stanford University Press.

Watts, Michael. 2004. "Resource Curse? Govermentality, Oil and Power in the Niger Delta, Nigeria." In *The Geopolitics of Resource Wars,* edited by Philippe Le Billon, 50–80. London: Routledge.

Watts, Michael. 2012. "A Tale of Two Gulfs: Life, Death and Dispossession along Two Oil Frontiers." *American Quarterly 64* (3): 437–467.

Zalik, Anna. 2011. "Labeling Oil, Reconstituting Governance: Legaloil.Com, the Global Memorandum of Understanding and Profiteering in the Niger Delta." Essay. In *Oil and Insurgency in the Niger Delta: Managing the Complex Politics of Petro-Violence,* edited by C. I. Obi and Siri Aas Rustad, 184–199. London: Zed Books.

# 5   Converting threats to power

## Methane extraction in Lake Kivu, Rwanda

*Kristin Doughty*

"Our #kivuwatt increases electricity access while solving critical risk." This tweet was sent on March 10, 2016 by an American company called Contour Global that designed, built, and operates Kivu Watt, an innovative plant that had recently begun extracting methane gas from Lake Kivu, Rwanda. Lake Kivu is a 2370-square kilometer lake that straddles the border of Rwanda and the Democratic Republic of Congo (DRC). The lake contains an estimated 60 billion cubic meters of dissolved methane gas, and an estimated 300 billion cubic meters of dissolved carbon dioxide. This dissolved gas has long been seen as a potential threat, a concern that was raised after a 1986 deadly eruption in a similar lake in Cameroon, Lake Nyos, asphyxiated approximately 1700 people. By comparison, Kivu has 1000 times more gas than Nyos, and an estimated two million people live in the Lake Kivu basin.

In January 2016, after more than 6 years of investment and building and to much global attention, Kivu Watt began methane extraction operations as the first industrial-scale gas-fueled power project in Rwanda and in the world. The methane extraction began providing 25 megawatts of electricity to Rwanda's national grid, an increase of 25 percent to overall capacity. Contour Global, which was granted a 25-year concession for Kivu's methane, planned to build another three barges to increase capacity to 100 megawatts. The government of Rwanda also granted a contract to a second American company, Symbion LLC, to build another power project extracting methane, with construction beginning in late 2016.

The day that tweet was sent, I was in Rwanda, in the town of Karongi (formerly Kibuye), where the Kivu Watt power plant is located, and where the extraction barge sits off shore, linked by a 13.5 kilometer pipeline submerged underwater. That morning, I visited the national Museum of the Environment, which had opened 9 months earlier (in July 2015), developed by the governmental Institute of National Museums in Rwanda. This new museum, with an explicit focus on "the environment through attention to energy," had a panel display devoted to methane extraction in Lake Kivu. The museum narrated the methane extraction project in a manner consistent with Contour Global's emphasis on increasing electricity while solving risk. In three languages—French, English, and Kinyarwanda—a panel called "Gas in Lake Kivu" explained, "The exploitation of this gas is important for several reasons. It would reduce the wood usage, it would avoid a limnic eruption, and presents an opportunity for economic development." The panel went on to draw a comparison to Cameroon's lakes, noting that "Various causes, such as earthquakes, landslides or volcanic activities can cause limnic eruptions, in which the gas content of the lake is released and in turn causes suffocation of animals and human beings." The panel closed by making explicit the link between extraction and risk reduction, explaining that, "In Rwanda, the exploitation of methane gas reduces the gas concentration of methane and carbon dioxide in the lake, turning it much safer."

Both of these examples—the tweet from Contour Global and the official framing in the museum panel—illustrate that the Rwandan government and corporate partners agree publicly that the methane extraction project has two aims: (1) to reduce dangerous levels of unstable gasses dissolved in the lake and thus prevent it from exploding, and (2) to provide much-needed power to meet increasing demand in energy-strapped countries (Rwanda and the DRC). The Kivu methane extraction project thus seeks to convert a potential natural disaster into a sustainable source of local energy in a region marked by histories of violence.

In this chapter, based on preliminary fieldwork with the Kivu methane extraction project, I argue that more attention needs to be paid to the kind of work this conversion narrative does to align a series of diverse actors and interests, and to the social, cultural, and political work that sustains this conversion. Further, I suggest that we must consider this methane extraction project a form of what Dominic Boyer has called "energopower," in which the management of populations is increasingly intertwined with the harnessing of energy (Boyer 2014). Doing so allows us to consider how energopower articulates with the Rwandan government's national unity and reconciliation policies to exacerbate or mitigate contemporary characterizations of Rwandan state power as authoritarian. This analytic frame sheds light on the imbrications of the cultural politics of energy and unity that are important to understanding socio-political dynamics across the African continent and elsewhere.

I begin by situating this project within recent anthropological research on energy. I then examine the construction of the discourse of risk of Lake Kivu, noting the shift from attention to political violence to a depoliticized "natural" lake. I then turn to more ethnographic exploration of the early phases of the Kivu methane extraction by examining how a change in the color of the lake demonstrates the multiple epistemologies that surround the extraction, and the stakes of the conversion work. I close by considering how this lake helps us to examine the interrelationships between efforts to control violence, people, nature, and energy.

## Anthropologies of energy

A growing body of scholarship argues that around the world, contemporary governance is inextricably tied to energy extraction, production, and use

(Boyer 2014; Mitchell 2011; Nader 2010). Anthropological work has long conceptualized natural resources used to generate energy—such as Kivu's methane—not as purely natural, but as produced through socio-cultural processes linked to local ontologies, complex webs of expertise, labor and global capitalism, and infrastructures (Ferry 2013; Rolston 2013; Strauss, Rupp, and Love 2013). Recent scholarship grounds analysis of these "resource materialities" (Richardson and Weszkalnys 2014) in the constitutive imbrications of nature and culture at the core of post-humanism and the new materialism (Cruikshank 2005; Helmreich 2011; Ingold 2012; Kohn 2013). One body of work explicitly examines how these resource extraction projects lead to "disasters" (Bond 2013; Cepek 2012; Kirsch 2001; Weszkalnys 2014), suggestive for the Rwandan case in which efforts to extract methane to avert a natural disaster (an eruption of deadly gas) also risk producing disaster through the extraction process itself. Other work equally relevant to this case examines how resource extraction projects become sites of struggle, generating conflict among stakeholders to access or control them (Ballard and Banks 2003; Coumans 2011; Golub 2014; Kirsch 2014), often leading to sustained violence (Adunbi 2015; Hoffman 2011, 2013; Kabamba 2013; Timsar 2015).

Much recent work on energy and resource extraction centers on oil, as ethnographers have shown how living "with" (Breglia 2013), "in the time of" (Leonard 2016; Limbert 2010), or "in" oil (Cepek forthcoming) shapes forms of sociality, culture, and politics of everyday life. Scholars argued a decade ago that oil is particularly crucial to understanding contemporary Africa (Barnes 2005; Watts 2004), given the increasing "scramble" for African oil (Frynas and Paulo 2007). Work emerging from West Africa and Angola examines how oil extraction assemblages reshape political–economics on the continent (Behrends, Reyna, and Schlee 2011; Guyer 2015; Leonard 2016; Soares de Oliveira 2007; Watts 2004) and how these governance and landscape transformations reshape people's life worlds (Adunbi 2011, 2013; Appel, Mason, and Watts 2015; Timsar 2015), including with deep inequalities (Appel 2012b; Reed 2009). Work in Africa illustrates the particular dynamics of offshore drilling in international waters (Appel 2012a; Campos-Serrano 2013; Chalfin 2015) and emphasizes corporations' extractive enclaves (Appel 2012a; Ferguson 2006). This recent scholarship builds on longstanding interest in resource extraction in Africa, particularly work that used the Copper Belt mines in Southern Africa as central for examining the impact of colonization and modernity in Africa (Ferguson 1999; Moore 1994). Other work examines the transformative impacts of electrification, and its imbrications with political-economic power (Degani 2013; Winther 2008). Work on extraction and electrification in Africa connects to broader questions about sovereignty, development, dependence, the politics of distribution, and future-making on the continent (Comaroff and Comaroff 2011; Ferguson 2015; Piot 2010).

Together, these insights point to the need to examine how the Lake Kivu methane extraction project shapes governance across private and public regimes, at national and international levels, and to examine how the particular materiality, ontologies, expertise, and infrastructures of methane extraction shape people's lives at intimate levels. I suggest we can examine the Kivu methane extraction project as a form of "energopower," in which management of populations is increasingly intertwined with the harnessing of energy (Boyer 2014). Energopower as a concept underscores how energy and populations are subject to parallel forms of technocratic management, and how biopolitics and energy politics logically presuppose each other. Energopower, Boyer notes, remains an open, even nebulous, concept worthy of additional clarification in the contemporary moment, particularly the relationship between energopower and biopower. Further research on Kivu's methane extraction, I hope, will shed light on energopower in Rwanda and perhaps outside. Does the management of methane parallel and precede a relatively autonomous bio-political project of managing the Rwandan people (*abaturage*), or is the population somehow "methanized" in and through this project? I suggest that to understand the ways Rwandan people are part of the methane assemblage requires examining how energopower in Rwanda articulates with other government rebuilding policies. How is energopower connected to other kinds of population control, both ideological and legal, but also sub-juridical and sub-ideological, to produce particular kinds of citizens in *Rwanda Rushya*, the new Rwanda? This frame, which I introduce here and will pursue in further fieldwork, allows us to ask how Rwandans recovering from the genocide experience official efforts to control violence and promote national unity as intertwined with energy agendas.

## Constructing and deconstructing the Risky Lake

Rwanda is a particularly productive site for examining connections between energy and national unity because of its unique combination of recent recovery from genocide with strong governance and sufficient political stability to pursue internationally supported cutting-edge scientific resource extraction. The 1990s were marred by extensive bloodshed in Rwanda that killed approx-imately one million people, and displaced hundreds of thousands of others. The decade was punctuated by a civil war from 1990–1993, a genocide in 1994 (a campaign of state-backed violence designed to eliminate the country's minority Tutsi population), and then years of periodic military reprisals and massacres (Des Forges 1999; Fujii 2009; D. Newbury 1998; Prunier 1995; Straus 2006). The Rwandan government's rebuilding efforts were initially lauded as exemplary in controlling violence and promoting economic growth, as Rwanda has so far successfully avoided overt internal political conflict, in contrast to neighboring DRC, Burundi, Uganda, and Kenya. In particular, the Rwandan government appeared to emphasize rule of law as central to its reconstruction, and rebuilt Rwanda in accordance with global best-practices in the realms of transparency, gender equality, and environmental standards. The government proved self-assured in working with international institutions such as the World Bank to secure loans, and the United Nation's International Criminal Tribunal for Rwanda to promote accountability for genocide. Rwanda's effort to extract the vast reserves of methane and carbon dioxide dissolved in Lake Kivu has received international attention and seems further proof of Rwanda's status as a diamond in the rough (Rosen 2015).

Interest in Lake Kivu's natural gases is longstanding both as a potential power source and for its risk of unpredictable release. A small Belgian-built power plant on the eastern shore began operating as early as 1963 (Doevenspeck 2007). Efforts to extract methane from Lake Kivu were halted by the 1994 genocide in Rwanda and the surrounding violence. Scientific interest in Kivu was reanimated in 2005 with publication of a paper by a team of European scientists which showed that,

> methane production within the sediment has recently increased, leading to a gas accumulation in the deep waters and consequently decreasing the heat input needed to trigger a devastating gas release. With the estimated current [methane] production, the gas concentrations could approach saturation within this century.
>
> (Schmid et al. 2005, 1)

That is, the authors argued that methane levels had increased dramatically between 1974 and 2004, and if they continued increasing at that rate would reach unsustainable levels by 2104, with an unpredictable degassing event possibly triggered by volcanic activity or earthquake (Schmid et al. 2005). At approximately this time, the Rwandan government developed a Memorandum of Understanding with the DRC outlining the terms of shared ownership of the methane resource in the lake. The Rwandan government also intensified pursuit of partnerships to operationalize methane extraction. In 2008 a small plant called KP1, jointly developed by the Rwandan government and a foreign firm, began operating intermittently.

By the late 2000s, articles began to appear in international and regional African media emphasizing the risks posed by Lake Kivu. The journal *Nature* in 2009 referred to Kivu as a "lakeful of trouble," both "dangerous and valuable" (Nayar 2009). It was subsequently called a "dangerous lake" (PRI 2012) or even a "deadly" (Mena Report 2014) or "killer lake" (Vltchek 2011), a "freshwater time bomb" (Gawlowisz 2016), with the gas itself described as a "hidden menace" (*Economist* 2016) or "toxic menace" (Baker 2016). How and why the discourse of the risk of "explosion" shifted from a focus on an ethnicized, radicalized people exploding against one another in the midst of Africa's World War, to a body of water—a seemingly neutral, natural, apolitical feature of the landscape—warrants further examination. That is, we must first probe and denaturalize the construction of risk, on which the conversion of threat to benefit is predicated.

A great deal of work has shown how seemingly natural threats, such as the 2011 tsunami in Sri Lanka (Choi 2015) or Hurricane Katrina (Adams 2013; Button 2010), are socio-political insofar as they are defined, analyzed, and controlled in relation to government policies, scientific expertise, and private interests at national and international levels—as are more obviously human-produced disasters like oil spills (Bond 2013), nuclear accidents such as in Bhopal (Fortun 2001), or urban fires and crime (Auyero and Swistun 2009; Klinenberg 2002). The ostensibly apolitical dimensions of techno-scientific projects in fact have deeply political and even violent effects (Mitchell 2002), perhaps especially when combined with the logics of capitalism and private enterprise. Considering the framing of Lake Kivu's risk in relation to energopower suggests the need for further examination into how, in a society defined by risk or emergency (Beck 2009), the Rwandan government might be using the discourse of threat and anticipation of a potential natural disaster to shape cross-border relations and promote national belonging through risk management and "disaster nationalism" (Choi 2015, 293).

Writing about the May 2016 official inauguration of the Kivu Watt extraction, presided over by President Kagame, *Time* magazine presented the prevailing view that, "[I]f released, this toxic combination could take the lives of the more than two million people living along the lake shore" (Baker, 2016). Yet in my preliminary work on this new project, I have heard very differing views on the actual risk of Kivu's methane from expert international scientists with intimate knowledge of the lake and the geoformations of the region. Some scholars say that even releasing the gas into the air, as is done routinely in Nyos and a second similar lake in Cameroon, would be preferable to the impacts of a one-off spontaneous escape of gas,[1] while others say that such venting would be "environmentally irresponsible" (Baker 2016). Other scientists say that if a spontaneous overturn occurred, it would destroy the ecology of the lake, but not the terrestrial population.[2] Some point out that even the extraction will not really keep pace with the methane creation, and therefore are skeptical that the process is indeed reducing risk.[3] Others claim that the bigger risk emerges from the extraction itself, if it disrupts the stability of the lake.[4] There is local knowledge of the lake's unpredictable danger—a missing boat, or grazing animals found inexplicably dead by the shore[5]—though it is unclear as yet how people living along the lake fit the methane extraction project into these ideas of risk, or how these risks are assessed in relation to other threats.

In contemporary Rwanda, probing definitions of risk is important because of the many parallel examples of how the governing regime strategically controls official definitions of key concepts for political ends. Specifically, scholars have pointed to how the government's master narrative of history and description of "the genocide" is central to justifying contemporary governance policies (Burnet 2012; Meierhenrich 2008; Vidal 2001). This official view of the genocide is a narrative of the violence of the 1990s with carefully constructed distinctions between criminal and prosecutable violence as differentiated from political and therefore justifiable violence, with resulting notions of culpability and victimhood. Scholars have further noted how Rwandan officials marshal "genocide credit" at the international level, playing on global guilt for non-intervention during the genocide to raise aid money and insulate questionable policies from external critique (Reyntjens 2005). Meanwhile, in the same month that Kivu Watt began its extraction, December 2015, Rwandan voters passed a referendum to amend the constitution to allow President Kagame to run for a third term. The Rwandan government dismissed international criticism of the referendum vote—in which the Rwandan Electoral Commission claimed 98 percent voter turnout, with 98.35 percent voting in favor of the proposed change—asserting that its citizens had spoken, and pointing to ongoing peace and stability.[6]

Further, there are ample reasons to be wary of nationalistic policies designed to manage risks by transforming threats into valuable resources. The post-genocide Rwandan government has used "managing risk" to justify military violence against alleged *génocidaires* in Rwanda as well as in neighboring DRC, where Rwanda's government has been accused of military incursions and massacres including with explicit links to control over natural resources (Lemarchand 2009a; OHCHR 2010; Prunier 2009; Reyntjens 2011). The Rwandan government has further used managing the risk of violence in the name of national unity and reconciliation to regulate people's speech and behavior—for example, through making public reference to ethnicity illegal and criminalizing language that can promote ill-will in a 2008 law against genocide ideology. These risk-management efforts, while perhaps laudable in intent, have censored civil society, eliminated political opposition, and promoted fear and coercion among ordinary citizens (Buckley-Zistel 2009; Chakravarty 2016; Ingelaere 2010; Lemarchand 2009b; Reyntjens 2005; Straus and Waldorf 2011; Thomson 2013).

Also at the level of official discourse of converting threat to benefit, there appear to be similarities between the government's framing of the Kivu methane project and other cultural-nationalistic solutions to post-genocide reconstruction. The Kivu methane project seems to be narrated as another example of using local Rwandan resources to solve domestic problems, much like the decision to draw on Rwandan customary law (through *gacaca* courts) instead of Western-style courts for trying hundreds of thousands of alleged genocide perpetrators (Chakravarty 2016; Clark 2010; Doughty 2016; Ingelaere 2009; Thomson 2011; Waldorf 2006). Like customary law, local energy is framed as benign and justified as more naturally Rwanda, and I suspect that the implementation of methane extraction, like customary law, involves logics of coercion and accompanying resistance that warrant further examination.

Further, the conversion of dangerous methane into positive energy can be seen as the latest instantiation of a broader pattern whereby the Rwandan government converts potential threats into resources, with unequal benefits. Official policies to convert threats into assets, such as transforming genocide perpetrators into free communal labor, purport to benefit the whole country but actually serve to consolidate the government's hold on power while further marginalizing vulnerable segments of the population (Ansoms 2011). This effect is likely to be heightened when disaster mitigation is pursued through for-profit enterprise, as Adams described in post-Katrina New Orleans

(Adams 2013), particularly given the ruling Rwandan Patriotic Front's (RPF) notoriously close connections with private business operations (Booth and Golooba-Mutebi 2012).

Indeed, the emphasis on threat reduction risks erasing the role of economic profit in the operation of this project, and ultimately on its impacts. The extraction is financed through loans from the African Development Bank, Emerging Africa Infrastructure Fund, Netherlands Development Finance Company, and the Belgian Investment Company for Developing countries (Kimenyi 2016). Contour Global's efforts are often described in nearly humanitarian terms, as a "noble undertaking" (Baker 2016) with little reference to the profit margins they are expected to gain, even as one involved person told me that it is indeed a "very profitable business" and Contour Global will likely recoup their investment within 5 to 6 years.[7] During that same week in March when the tweet was sent, a person closely involved in the project commented to me, "I am surprised to hear you saying that our job is minimizing risk, since I've been thinking our job is optimizing the resource."[8] By contrast, Lake Kivu Monitoring Program staff were quick to emphasize that efficiency was the last of their three objectives, which were to: "1. Ensure public safety; 2. Preserve the Ecosystem; 3. Maximize socio-economic benefit" (LKMP poster, in author's possession).

The Lake Kivu area is of high interest to the Rwandan government for national security reasons, given the international border with the DRC, and also for tourism, given both the lake's beauty and the proximity to the Parc National des Volcans, home to globally renowned mountain gorillas and volcanoes. The methane extraction project is part of Rwanda's wider move to "go green," including policies such as banning plastic bags and trash burning, and supporting investment in innovative alternative energy projects, including biofuel, geothermal, and solar, even as traditional mining and hydropower are on the rise. Overall, I suggest that the analytic frame of energopower can help bring into relief the socio-political work of converting a threat to a benefit and the ways this narrative aligns particular assemblages of actors. It can draw our attention to the constellation of corporate, state, and multinational assemblages that shape governance, and to the ways people experience multiple domains of power in interconnected ways.

## Conversions and transformations: lake color change

Contour Global, the company behind the current methane extraction, was founded in 2005, based in New York, and currently has 61 power plants across the world, spanning 20 countries and three continents, including both renewable and non-renewable sources, and conventional and innovative strategies.[9] Their investment in Rwanda was part of a wider African regional strategy where they "believe in the power of energy to drive development, increasing stability for citizens and businesses."[10] Their approach, their promotional materials explain, emphasizes "developing custom solutions from local resources. Project

KivuWatt is a prime example of this innovative thinking in action."[11] The in-country management and engineering staff were a multinational mixture, with a few Rwandan engineers and office staff, with other Rwandans hired mostly for construction labor.

The KivuWatt project took shape over more than 4 years, and required extensive infrastructure transformations in western Rwanda. The process for figuring out the extraction was complex. One of the unique elements of Lake Kivu is that not only does it contain high levels of methane, but this methane exists within permanently stratified layers of water (Pasche et al. 2011). The top layer, to a depth of 60 meters, is paradoxically "nutrient poor" and contains the fish that drive the active local fishing industry on the lake. The "resource zone," which is at a depth between approximately 250 meters to 492 meters, "is the one containing the much important concentration of gas and is the zone from which gas is extracted" (LKMP poster, in author's possession). Preserving this stratification is crucial to preventing gas release.

The Contour Global team developed a strategy for extraction based on what was being used at the pilot KP1 plant, but scaled up. They considered the technology highly proprietary, and went to great lengths to shield it from competitors.[12] On the KivuWatt barge, the methane-rich water is pulled up from the deep resource zone. The methane is removed from the gas-rich water above the surface. The degassed waste water is then reinjected into the lake in a middle layer of the lake water, but not as deep as the gas-layer. The methane is carried through a submerged pipeline to the new onshore power plant. There it is immediately converted into electricity, and delivered onto the national grid. In order to achieve this, the Rwandan government provided new roads and power lines to assist Contour Global's infrastructural needs. Contour Global built a marine landing site, at which the barge was constructed and then launched. They built a power plant, tucked behind the hills several kilometers out of town. It was, as one Contour Global employee described, "A very challenging technical process" with "lots of starts and stops."[13] A top Kivu Watt manager explained that during the building phase, the project manager was "practically crying every day," and it took a great deal of perseverance, and failures and re-dos, before extraction was successful. The barge was launched in mid-2015, and began pilot operations in late 2015, leading up to full-capacity power generation by January 2016. Anecdotally, people in Kigali reported that power quickly became more stable at that point, with fewer spontaneous cuts.[14]

In April 2016, less than 4 months after extraction operations had begun in earnest, Lake Kivu suddenly changed color, virtually overnight. From afar, the lake looked turquoise as distinct from its usual opaque blue-black, while up close it appeared milky white. The change was equally noticeable to the naked eye and to satellite images, and was captured by vacationers alongside the lake. I learned of the dramatic color change from a colleague who emailed me to say that I should stay appraised, as people in many different positions with respect to the methane extraction and the lake were all trying to figure out what it meant. As a Rwandan news agency reported, the color change "caused panic" and "people around the lake are spreading myths and fairy tales which continue to create a sense of fear" (Sebuharara 2016). These newspaper quotes, of course, suggest a class disjuncture between educated urban Rwandan reporters and rural Rwandan fishermen and farmers, as well as an implied distinction between false myths and true science. Relevant for my discussion here was how this color change put into relief the stakes of the narrative of conversion of threat into benefit, and how it risked being undone.

The Contour Global team explained that they were alarmed by the color change and how people would interpret it. Two executives at Contour Global with whom I spoke separately about the lake's color change both indicated that Contour Global took a proactive approach, as part of their ongoing efforts to emphasize community engagement, core to their foundational principles. One described, "When we saw the lake turning green [then white], we were immediately alarmed" because "we want to tell people, the parties, that it's not related to the extraction."[15] One was quick to emphasize to me that "There is a fear of the lake among the community."[16] Staff told me that they were even contacted by phone and email by NASA scientists, who had noticed the color change in satellite images, and wanted an explanation.[17]

Contour Global indicated that in addition to consulting with their own in-house experts, they reached out immediately to staff at the Lake Kivu Monitoring Program (LKMP), the government agency that monitors the environmental risks of the methane extraction. The LKMP was initially established in 2006, and began operations in 2008, with the onset of the first pilot extraction plant (KP1). In 2011, it moved to the Energy and Water Sanitation Authority (EWSA). Now it is overseen by the development branch of the governmental Rwanda Energy Group (REG), within the Energy Development Corporation Limited (EDCL). Its headquarters were in Kigali, with a second office on the northern shore of the lake, in Rubavu (formerly Gisenyi) near the DRC border. Staff were predominantly Rwandan, with a few international employees on 1-long contracts, including a Dutch engineer responsible for on-site evaluations of the extraction, and a Belgian biologist overseeing the laboratory work. LKMP concentrated primarily on evaluating the physico-chemical properties of the lake, including the impacts on fish populations, and conducting on-plant, near-plant, and lake-wide monitoring near the KP1 plant. Specifically, they monitor the wastewater that is reinjected, because it could impact the nutrient load of the upper layer of the lake, adding nutrients from the lower levels that could promote plankton blooms that paradoxically destroy the fish populations.

LKMP staff told me that this was not the first time that the color had changed; lesser events occurred in 2005 and 2013. Yet, they indicated that this time was more important because, "It was the first time we were contacted by the media," and, "It was a difficult time to explain."[18] LKMP staff members were the point people working between the various constituencies, and their response points to the processes of commensuration collaborators engaged in as they sought to manage disparate understandings of this "natural" event. LKMP staff conducted water testing to determine the conductivity, temperature, and depth profiles (CTD) of the water before and after at their shore-side laboratory, and reached out to international consultants for help with analysis.

The international scientists who were consulting with LKMP provided highly specialized analyses of what was happening with the water. These scientists, mostly based in the United States or Europe though some also in the DRC, compared this color change to similar examples in Lake Geneva, Switzerland and Lake Ontario, on the Canada/US border. They described the phenomenon as calcite precipitation, saying: "Due to algal blooms, pH is shifted to higher values, i.e. water becomes alkaline. As a consequence, the carbonate equilibrium is shifted towards $CO_3$—; obviously concentrations are sufficient to overcome the solubility product of $CaCO_3$; i.e. calcite is precipitated" (email, May 2016). This evidence was presented as, and interpreted as, factual as contrasted with the "fairy tales" in which rural residents were imagined to believe. One senior expert on the lake explicitly addressed the question of whether the color change could be related to the extraction—or, more specifically, the reinjection of degassed water that LKMP was monitoring— and said very clearly that the color change was, "Definitely not related to deep water re-injection which so far is behaving as modeled, in my opinion" (email May 2016). Another wrote, "NO. The reinjection is deep, far away from the scene of the action here" (email May 2016). But one scientist pointed out that managing communication and perceptions about these issues was extremely important because, "Even a detail can create rumors and panics" (personal correspondence, May 2016).

When I asked who, precisely, was panicking, responses were unclear. Some suggested that ordinary people were not so concerned, while it was more the journalists and educated people who knew something, but not enough.[19] Colleagues in Rwanda indicated that local people were concerned, yet there was no systematic effort to formally gather their views. Lakeside residents wondered about whether this signaled a forthcoming gas release, whether it was related to recent activity in adjacent volcanoes, or whether it was linked to the methane extraction. Many of the people noticing the lake color change were fishermen, as Lake Kivu has an active fishing industry based on stock estimates of approximately 5,000 tons of isambaza, which have held mostly steady over the past 20 years (LKMP paper, 2016, in author's possession). Many of these fishermen were already upset over a recent debate about access to the lake.[20] Fishermen were told they had to maintain a 1-kilometer distance from any Contour Global project infrastructure. Given that the pipeline extending from the barge to shore was 13.5 kilometers long, this blocked off a large portion of the lake. It effectively served as a gate that blocked fishermen, in their traditional dug-out canoe boats, from getting out to the sections of the lake where they typically fished. Contour Global maintained that this was not their policy, but rather that of Rwanda's marine security and border patrol. This incident had brought together parties from LKMP, Contour Global, and fishing cooperatives in an effort to discuss and resolve it, but the conflict remained fresh.

The example of the lake color change, which may have compounded anger over restrictions in fishing access, suggests the importance for ongoing research into perceptions of Rwandans affected by this project, and others like it, in order to better understand the imbrications of the cultural politics of energy and unity. Emerging work shows how energy projects in post-conflict contexts serve as sites for ongoing contestation over key political concepts that are crucially in play in the wake of violence, i.e., through hydro-fracking in South Africa (Green 2014) or biofuel in Sierra Leone (Maconachie and Fortin 2013). Scholars have repeatedly shown that in post-conflict contexts emerging from high-stakes fractures that led to intense violence, agreement on the meaning of reconciliation, justice, threat, or the public good cannot be presupposed, but people's interpretations vary widely and often differ from official versions (Abramowitz 2014; Theidon 2013; Wilson 2001). That is, the post-genocide context amplifies our need to examine, rather than assume, how people understand key concepts such as risk or benefit, and how they relate to people's broader ideas of justice, citizenship, and imagined futures.

In addition, people's perceptions of methane extraction, and episodes like Kivu's recent color change, are likely bound up with their experiences of the genocide. Research has shown clearly elsewhere that as people navigate the dramatic transformations in landscape that energy and natural resource projects bring, they produce new forms of political association and belonging, often with respect to indigeneity or ethnicity (Appel, Mason, and Watts 2015; Breglia 2013; Muehlmann 2013). Work on social memory has shown that landscapes serve as "sites of memory" (Nora 1989), in which people's perceptions of the past shape group belonging and views of state policies in the present (Connerton 1989; Halbwachs 1980), especially in Rwanda (Lemarchand 1998; Malkki 1995; C. Newbury 1998). Despite the official erasure of ethnicity in Rwanda, divisions among Rwandans have not disappeared, but rather people's understandings of the genocide and civil war, and associated notions of guilt and victimhood, heavily influence how they relate to one another and how they engage with the state (Burnet 2012; Eramian 2014a, 2014b; Hilker 2009, 2012; Sommers 2012). We must ask, then, how do Rwandans understand and experience transformations in the environment brought by the Kivu methane project—new barges, new infrastructures, increased pollution along the lake shores, and reduced fishing—in relation to the politics of victimhood and "genocide citizenship" (Doughty 2016)?

In early fieldwork, people spontaneously narrated the lake and its infrastructure to me as situated within Rwanda's political history in ways that illustrated the imbrications of the project in local histories, and highlight the stark variations among various actors' framing of the project. For example, when I commented to one colleague in Kigali about the poor condition of the road alongside one portion of the lake, he exclaimed, "Still? It has been 22 years!" He went on, "The genocide began on April 6 and I was there on April 1 and they were just starting to put out the equipment to flatten the road."[21] Another man in Kigali, commenting on how people in Kibuye feel ownership of the lake and resist external regulation and interference on it explained that, "In Habyarimana's time, when the road was being built to Kibuye, people there were saying, 'They are going to take our lake.' Even at that time, people thought of it as 'our lake.' "[22] These provocations emphasize the need for further consideration of contextualized, historicized views of the lake and the methane extraction project to understand its impacts on the people purported to benefit.

This episode of Kivu changing color shows how the conversion of risk into benefit brought together particular alliances of actors across different epistemologies and forms of expertise. It highlights the labor of meaning-making in these collaborations—the efforts to render proportional diverse understandings of one "natural" event, across different regimes of knowledge, authority and expertise. Further, it shows how the conversion had to be actively sustained—as this color change risked undoing the conversion of threat to benefit, or reversing people's perception of its direction from benefit back to risk if they came to believe the color change was caused by extraction. The final word was—at least for now, according to LKMP and international scientists, as printed in a Rwandan news source—that the lake change was "a natural, safe, and reversible phenomenon," "not caused by the current extraction of methane gas nor by recent seismic and volcanic activities," and "Lake Kivu has not yet been affected by the global warming," (Sebuharara 2016). Yet, we do well to remember the imbrications of natural, cultural, and political dimensions of this phenomenon. That is, even changes in the color of a massive lake must be understood as related to the workings of state power, and people's experience of post-genocide recovery. Especially given the ways scientists, corporations, and governments are working together to extract methane, it seems important to keep asking how these collaborations produce and sustain energopower in contemporary Rwanda, and what the nature of energopower is, as distinct from but imbricated with other state-backed efforts of national unity and development.

## Conclusion

Around the world, natural resource extraction is often used to drive development in the wake of both political violence and natural disaster. At the same time, the extraction of resources for energy is paradoxically linked to conflict, whether in the Middle East or Niger Delta, as well as to the growing incidence of natural disasters through anthropogenic climate change. Despite these seemingly obvious connections, it remains unclear how official state-backed efforts to control violence, people, nature, and energy are intertwined. While this volatile lake is unique to the Rwanda–DRC border, it underscores the centrality of the techno-scientific projects that enable energy extraction projects that are of critical importance to understanding energopower in contemporary Africa, and of the collaborations at their heart.

With the growing importance of the Lake Kivu methane extraction and its promise to solve two problems simultaneously by converting threats into power, there remains a striking absence of attention to several critical dimensions of this project, including the interconnections with other forms of governance and resource extraction in Rwanda and Africa more widely, and the perspectives of people affected by it. These lacunae promote a view of the project as politically and ethically neutral, even as energy policies are anything but benign (Behrends, Reyna, and Schlee 2011; Nader 2010). I have used this chapter, based on preliminary research during the early months of methane extraction, to argue for the need to incorporate these experiential and analytic perspectives more foundationally into our understanding of this energy project, in order to understand how people are governed through this extraction. Further fieldwork on this project could shed light on the nature of energopower, and the extent to which people are governed through methane or adjacent to it, as with other extractive resource enclaves in Africa. More broadly, this methane project and people's experiences of it might point to the limits of the post-genocide Rwandan government's strategies of legitimation.

## Acknowledgments

Research for this project is funded by grants from the Wenner-Gren Foundation for Anthropological Research and the National Science Foundation, with additional financial support from the University of Rochester's Humanities Center Fellowship. Research is covered by Rwandan research clearance #MINEDUC/S&T/369/2016 and University of Rochester's human subjects review board. Thanks go to colleagues in Rwanda, including at the LKMP, University of Rwanda, and at Contour Global for sharing their time. Thanks to Peter Castro for comments on an earlier version of this paper which was presented at the Society for Cultural Anthropology meetings in Ithaca, NY in May 2016, and to Michael Degani and Lori Leonard for comments on this chapter.

## Notes

1  Interview. March 31, 2016.
2  Interview. April 7, 2016.
3  Interviews. March 2016.
4  Interviews. February 2016.
5  Interview. March 2016.
6  By contrast, violence in neighboring Burundi escalated in late 2015 in response to president Nkurunziza's attempt to secure a third term in office. Many experts claim this descent has disturbing parallels to pre-genocide Rwanda.
7  Interview. March 2016.
8  Interview. March 2016.
9  www.contourglobal.com/company. Accessed July 15, 2016.
10  www.contourglobal.com/region/africa. Accessed July 15, 2016.
11  www.contourglobal.com/kivuwatt. Accessed July 15, 2016.

12  Interview. September 2016.
13  Interview. April 2016.
14  Interviews. September 2016.
15  Interview. May 2016.
16  Interview. May 2016.
17  Interview. September 2016.
18  Interview. May 2016.
19  Interview. September 2016.
20  Interview. March 2016.
21  Interview. September 2016.
22  Interview. September 2016.

# References

Abramowitz, Sharon Alane. 2014. *Searching for Normal in the Wake of the Liberian War.* Philadelphia: University of Pennsylvania Press.

Adams, Vincanne. 2013. *Markets of Sorrow, Labors of Faith.* Durham, NC: Duke University Press.

Adunbi, Omolade. 2011. "Oil and the Production of Competing Subjectivities in Nigeria: 'Platforms of Possibilities' and Pipelines of Conflict.'" *African Studies Review* 54 (3): 101–120.

Adunbi, Omolade. 2013. "Mythic Oil: Resources, Belonging and the Politics of Claim-Making Among Ìlàje Yorùbá of Nigeria." *Africa, 83* (2): 293–313.

Adunbi, Omolade. 2015. *Oil Wealth and Insurgency in Nigeria.* Bloomington: Indiana University Press.

Ansoms, An. 2011. "Rwanda's Post-Genocide Economic Reconstruction: The Mismatch between Elite Ambitions and Rural Realities." In *Remaking Rwanda: State Building and Human Rights after Mass Violence,* edited by Scott Straus and Lars Waldorf, 240–251. Madison: University of Wisconson Press.

Appel, Hannah. 2012a. "Offshore Work: Oil, Modularity, and the How of Capitalism in Equatorial Guinea." *American Ethnologist 39* (4): 692–709.

Appel, Hannah. 2012b. "Walls and White Elephants: Oil Extraction, Responsibility, and Infrastructural Violence in Equatorial Guinea." *Ethnography 13* (4): 439–465.

Appel, Hannah, Arthur Mason, and Michael Watts, eds. 2015. *Subterranean Estates: Lifeworlds of Oil and Gas.* Ithaca: Cornell University Press.

Auyero, Javier, and Debora Alejandra Swistun. 2009. *Flammable: Environmental Suffering in An Argentine Shantytown.* Oxford: Oxford University Press.

Baker, Aryn. 2016. "How Rwanda Turned a Toxic Menace Into a Source of Power." *Time Magazine,* May 17. http://time.com/4338310/rwanda-kivuwatt-methane-lake-kivu/?xid=newsletter-brief.

Ballard, Chris, and Glenn Banks. 2003. "Resource Wars: The Anthropology of Mining." *Annual Review of Anthropology 32*: 287–313.

Barnes, Sandra T. 2005. "Global Flows: Terror, Oil, and Strategic Philanthropy." *African Studies Review 48* (1): 1–23.

Beck, Ulrich. 2009. *World at Risk.* Translated by Ciarin Cronin. Malden: Polity Press.

Behrends, Andra, Stephen P. Reyna, and Günter Schlee, eds. 2011. *Crude Domination: An Anthropology of Oil.* New York: Berghahn Books.

Bond, David. 2013. "Governing Disaster: The Political Life of the Environment during the BP Oil Spill." *Cultural Anthropology 28* (4): 694–715.

Booth, David, and Frederick Golooba-Mutebi. 2012. "Developmental Patrimonialism? The Case of Rwanda." *African Affairs 111* (444): 379–403.

Boyer, Dominic. 2014. "Energopower: An Introduction." *Anthropological Quarterly 87* (2): 309–334.

Breglia, Lisa. 2013. *Living with Oil: Promises, Peaks, and Declines on Mexico's Gulf Coast.* Austin: University of Texas Press.

Buckley-Zistel, Susanne. 2009. "We are Pretending Peace: Local Memory and the Absence of Social Transformation and Reconciliation in Rwanda." In *After Genocide: Transitional Justice, Post-Conflict Reconstruction and Reconciliation in Rwanda and Beyond*, edited by Phil Clark and Zachary D. Kaufman, 125–144. New York: Columbia University Press.

Burnet, Jennie E. 2012. *Genocide Lives in Us: Women, Memory, and Silence in Rwanda.* Madison: University of Wisconsin Press.

Button, Gregory. 2010. *Disaster Culture: Knowledge and Uncertainty in the Wake of Human and Environmental Catastrophe.* Walnut Creek: Left Coast Press.

Campos-Serrano, Alicia. 2013. "Extraction Offshore, Politics Inshore, and the Role of the State in Equatorial Guinea." *Africa 83* (2): 314–339.

Cepek, Michael. 2012. "The Loss of Oil: Constituting Disaster in Amazonian Ecuador." *The Journal of Latin American and Caribbean Anthropology 17* (3): 393–412.

Cepek, Michael. Forthcoming. *Life in Oil: Surviving Disaster in the Petroleum Fields of Amazonia.* Austin: University of Texas Press.

Chakravarty, Anuradha. 2016. *Investing in Authoritarian Rule: Punishment and Patrongage in Rwanda's Gacaca Courts for Genocide Crimes.* Cambridge: Cambridge University Press.

Chalfin, Brenda. 2015. "Governing Offshore Oil: Mapping Maritime Political Space in Ghana and the Western Gulf of Guinea." *South Atlantic Quarterly 114* (1): 101–118.

Choi, Vivian Y. 2015. "Anticipatory States: Tsunami, War, and Insecurity in Sri Lanka." *Cultural Anthropology 30* (2): 286–309.

Clark, Phil. 2010. *The Gacaca Courts, Post-Genocide Justice and Reconciliation in Rwanda: Justice without Lawyers.* Cambridge: Cambridge University Press.

Comaroff, Jean, and John L. Comaroff. 2011. *Theory from the South: Or, How Euro-America is Evolving Toward Africa.* New York: Routledge.

Connerton, Paul. 1989. *How Societies Remember.* Cambridge: Cambridge University Press.

Coumans, Catherine. 2011. "Occupying Spaces Created by Conflict: Anthropologists, Development NGOs, Responsible Investment, and Mining." *Current Anthropology 52* (S3): S29–S43.

Cruikshank, Julie. 2005. *Do Glaciers Listen? Local Knowledge, Colonial Encounters, and Social Imagination.* Toronto: UBC Press.

Degani, Michael. 2013. "Emergency Power: Time, Ethics, and Electricity in Postsocialist Tanzania." In *Cultures of Energy: Power, Practices, Technologies*, edited by Sarah Strauss, Stephanie Rupp, and Thomas Love, 177–192. Walnut Creek: Left Coast Press.

Des Forges, Alison Liebhofsky. 1999. *Leave None to Tell the Story: Genocide in Rwanda.* New York: Human Rights Watch.

Doevenspeck, Martin. 2007. "Lake Kivu's Methane Gas: Natural Risk, or Source of Energy and Political Security." *Africa Spectrum 42* (1), 95–110.

Doughty, Kristin Connor. 2016. *Remediation in Rwanda: Grassroots Legal Forums.* Philadelphia: University of Pennsylvania Press.

*Economist*. 2016. "What Lies Beneath: Exploiting a Hidden Menace." *The Economist*. May 12. www.economist.com/news/middle-east-and-africa/21694554-exploiting-hidden-menace-what-lies-beneath.

Eramian, Laura. 2014a. "Ethnicity without Labels? Ambiguity and Excess in 'Postethnic' Rwanda." *Focaal: Journal of Global and Historical Anthropology* 70: 96–109. doi: http://dx.doi.org/10.3167/fcl.2014.700108.

Eramian, Laura. 2014b. "Personhood, Violence, and the Moral Work of Memory in Contemporary Rwanda." *International Journal of Conflict and Violence* 8 (1): 16–29.

Ferguson, James. 1999. *Expectations of Modernity: Myths and Meanings of Urban Life on the Zambian Copperbelt*. Berkeley: University of California Press.

Ferguson, James. 2006. *Global Shadows: Africa in the Neoliberal World Order*. Durham: Duke University Press.

Ferguson, James. 2015. *Give a Man a Fish: Reflections on the New Politics of Distribution*. Durham: Duke University Press.

Ferry, Elizabeth Emma. 2013. *Minerals, Collecting, and Value across the US-Mexico Border*. Bloomington: Indiana University Press.

Fortun, Kim. 2001. *Advocacy After Bhopal: Environmentalism, Disaster, New Global Orders*. Chicago: University of Chicago Press.

Frynas, Jedrzej George, and Manuel Paulo. 2007. "A New Scramble for African Oil? Historical, Political, and Business Perspectives." *African Affairs* 106 (423): 229–251.

Fujii, Lee Ann. 2009. *Killing Neighbors: Webs of Violence in Rwanda*. Ithaca: Cornell University Press.

Gawlowicz, Susan. 2016. "Volatile Gas Could Turn Rwandan Lake into a Freshwater Time Bomb." *University News*. November 6. Rochester: Rochester Institute of Technology. www.rit.edu/news/story.php?id=47155.

Golub, Alex. 2014. *Leviathans at the Gold Mine: Creating Indigenous and Corporate Actors in Papua New Guinea*. Durham, NC: Duke University Press.

Green, Lesley. 2014. Fracking, Oikos and Omics in the Karoo: Reimagining South Africa's Reparative Energy Politics. Conference paper presented at *The Thousand Names of Gaia: From the Anthropocene to the Age of Earth*, Rio de Janiero, Brazil, September.

Guyer, Jane I. 2015. "Oil Assemblages and the Production of Confusion: Price Fluctuations in Two West African Oil-Producing Economies." In *Subterranean Estates: Life Worlds of Oil and Gas*, edited by Hannah Appel, Arthur Mason, and Michael Watts, 237–252. Ithaca: Cornell University Press.

Halbwachs, Maurice. 1980. *The Collective Memory*. Translated by Francis J. Ditter Jr. and Vida Yazdi Ditter. New York: Harper and Row.

Helmreich, Stefan. 2011. "Nature/Culture/Seawater." *American Anthropologist 113* (1): 132–144.

Hilker, Lyndsay McLean. 2009. "Everyday Ethnicities: Identity and Reconciliation among Rwandan Youth." *Journal of Genocide Research* 11 (1): 81–100.

Hilker, Lyndsay McLean. 2012. "Rwanda's 'Hutsi': Intersections of Ethnicity and Violence in the Lives of Youth of 'Mixed' Heritage." *Identities: Global Studies in Culture and Power* 19 (2): 229–247.

Hoffman, Daniel. 2011. "Violence, Just in Time: Work and War in Contemporary West Africa." *Cultural Anthropology* 26 (1): 34–57.

Hoffman, Daniel. 2013. "Corpus: Mining the Border." Photo essay. *Cultural Anthropology*. https://culanth.org/photo_essays/1-corpus-mining-the-border.

Ingelaere, Bert. 2009. " 'Does the Truth Pass across the Fire without Burning?' Locating the Short Circuit in Rwanda's Gacaca Courts." *Journal of Modern African Studies* 47 (4): 507–528.

Ingelaere, Bert. 2010. "Peasants, Power and Ethnicity: A Bottom-Up Perspective on Rwanda's Political Transition." *African Affairs* 109 (435): 273–292.

Ingold, Tim. 2012. "Toward an Ecology of Materials." *Annual Review of Anthropology* 41: (427–442).

Kabamba, Patience. 2013. *Business of Civil War: New Forms of Life in the Debris of the Democratic Republic of Congo.* Oxford: CODESRIA.

Kimenyi, Brian. 2016. "Kagame Launches Kivu-Watt Power Plant." *New Times*, May 16.

Kirsch, Stuart. 2001."Lost Worlds: Environmental Disaster, 'Culture Loss,' and the Law." *Current Anthropology* 42 (2): 167–198.

Kirsch, Stuart. 2014. *Mining Capitalism: The Relationship Between Corporations and their Critics.* Berkeley: University of California Press.

Klinenberg, Eric. 2002. *Heat Wave: A Social Autopsy of Disaster in Chicago.* Chicago: University of Chicago Press.

Kohn, Eduardo. 2013. *How Forests Think: Toward an Anthropology Beyond the Human.* Berkeley: University of California Press.

Lemarchand, René. 1998. "Genocide in the Great Lakes: Which Genocide? Whose Genocide?" *African Studies Review* 41 (1): 3–16.

Lemarchand, René. 2009a. *The Dynamics of Violence in Central Africa.* Philadelphia: University of Pennsylvania Press.

Lemarchand, René. 2009b. "The Politics of Memory in Post-Genocide Rwanda." In *After Genocide: Transitional Justice, Post-Conflict Reconstruction and Reconciliation in Rwanda and Beyond*, edited by Phil Clark and Zachary D. Kaufman, 65–76. New York: Columbia University Press.

Leonard, Lori. 2016. *Life in the Time of Oil: A Pipeline and Poverty in Chad.* Bloomington: Indiana University Press.

Limbert, Mandana E. 2010. *In the Time of Oil: Piety, Memory and Social Life in an Omani Town.* Stanford: Stanford University Press.

Maconachie, Roy, and Elizabeth Fortin. 2013. " 'New Agriculture' for Sustainable Development? Biofuels and Agrarian Change in Post-war Sierra Leone." *Journal of Modern African Studies* 51 (2): 249–277.

Malkki, Liisa Helena. 1995. *Purity and Exile: Violence, Memory, and National Cosmology Among Hutu Refugees in Tanzania.* Chicago: University of Chicago Press.

Meierhenrich, Jens. 2008. "The Transformation of Lieux de Memoire: The Nyabarongo River in Rwanda, 1992–2009." *Anthropology Today* 25 (5): 13–19.

Mena Report. 2014. "Lake Kivu's Deadly Methane Becomes Source of Future Power." Amman, Jordan: Al Bawaba.

Mitchell, Timothy. 2002. *Rule of Experts: Egypt, Technopolitics, Modernity.* Berkeley: University of California Press.

Mitchell, Timothy. 2011. *Carbon Democracy: Political Power in the Age of Oil.* New York: Verso.

Moore, Sally Falk. 1994. *Anthropology and Africa: Changing Perspectives on a Changing Scene.* Charlottesville: University Press of Virginia.

Muehlmann, Shaylih. 2013. *Where the River Ends: Contested Indigeneity in the Mexican Colorado Delta.* Durham, NC: Duke University Press.

Nader, Laura, ed. 2010. *The Energy Reader.* Malden: Wiley-Blackwell.

Nayar, Anjali. 2009. "A Lakeful of Trouble." *Nature* 460 (7253): 321–323. www.nature. com/news/2009/090715/full/460321a.html.

Newbury, Catherine. 1998. "Ethnicity and the Politics of History in Rwanda." *Africa Today* 45 (1): 7–24.

Newbury, David. 1998. "Understanding Genocide." *African Studies Review* 41 (1): 73–97.

Nora, Pierre. 1989. "Between Memory and History: Les Lieux de Memoire." *Representations* 26, Special Issue: Memory and Counter-Memory (Spring): 7–24.

Pasche, Natache, Martin Schmid, Francisco Vazquez, Carsten J. Schubert, Alfred Wüest, John D. Kessler, Mary A. Pack, William S. Reeburgh, and Helmut Bürgman. 2011. "Methane Sources and Sinks in Lake Kivu." *Journal of Geophysical Research* 116: 1–16.

Piot, Charles. 2010. *Nostalgia for the Future: West Africa After the Cold War*. Chicago: University of Chicago Press.

PRI. 2012. "Rwanda Turning to a Dangerous Lake to Secure a More Independent Energy Future." PRI's The World, January 5. https://nortonsafe.search.ask.com/web?q=Rwanda+Turning+to+a+Dangerous+Lake+to+Secure+a+More+Independent+Energy+Future&o=APN11918&prt=NS&chn=1000&geo=US&ver=22&locale=en_US&guid=F005584A-498B-45D9-9ED9-C289ED772DE&tpr=111&gct=sb&qsrc=2869&doi=2016-10-19.

Prunier, Gérard. 1995. *The Rwanda Crisis: History of a Genocide*. New York: Columbia University Press.

Prunier, Gérard. 2009. *Africa's World War*. Oxford: Oxford University Press.

Reed, Kristin. 2009. *Crude Existence: Environment and the Politics of Oil in Northern Angola*. Berkeley: University of California Press.

Reyntjens, Filip. 2005. "Rwanda, Ten Years On: From Genocide to Dictatorship." In *The Political Economy of the Great Lakes Region in Africa*, edited by S Marysse and Filip Reyntjens, 15–47. New York: Palgrave MacMillan.

Reyntjens, Filip. 2011. "Waging (Civil) War Abroad: Rwanda and the DRC." In *Remaking Rwanda: State Building and Human Rights after Mass Violence*, edited by Scott Straus and Lars Waldorf, 132–151. Madison: University of Wisconsin Press.

Richardson, Tanya, and Gisa Weszkalnys. 2014. "Introduction: Resource Materialities." *Anthropological Quarterly* 87 (1): 5–30.

Rolston, Jessica Smith. 2013. "The Politics of Pits and the Materiality of Mine Labor: Making Natural Resources in the American West." *American Anthropologist* 115 (4): 582–594.

Rosen, Jonathan W. 2015. "Lake Kivu's Great Gas Gamble." *MIT Technology Review*, April 16. www.technologyreview.com/s/536656/lake-kivus-great-gas-gamble/.

Schmid, Martin, Maurice Hawbwachs, Bernard Wehrli, and Alfred Wüest. 2005. "Weak Mixing in Lake Kivu: New Insights Indicate Increasing Risk of Uncontrolled Gas Eruption." *Geochemistry, Geophysics, Geosystems* 6 (7): 1–11.

Sebuharara, Sylidio. 2016. Why is Lake Kivu Changing Colours? KT Press, April 26. http://ktpress.rw/2016/04/why-is-lake-kivu-changing-colours/.

Soares de Oliveira, Ricardo. 2007. *Oil and Politics in the Gulf of Guinea*. New York: Columbia University Press.

Sommers, Marc. 2012. *Stuck: Rwandan Youth and the Struggle for Adulthood*. Atlanta: University of Georgia Press.

Straus, Scott. 2006. *The Order of Genocide: Race, Power, and War in Rwanda*. Ithaca: Cornell University Press.

Straus, Scott, and Lars Waldorf, eds. 2011. *Remaking Rwanda: State Building and Human Rights after Mass Violence*. Madison: University of Wisconsin Press.

Strauss, Sarah, Stephanie Rupp, and Thomas Love, eds. 2013. *Cultures of Energy*. Walnut Creek: Left Coast Press.

Theidon, Kimberly. 2013. *Intimate Enemies: Violence and Reconciliation in Peru*. Philadelphia: University of Pennsylvania Press.

Thomson, Susan. 2011. "The Darker Side of Transitional Justice: The Power Dynamics behind Rwanda's Gacaca Courts." *Africa 81* (3): 373–390.

Thomson, Susan. 2013. *Whispering Truth to Power: Everyday Resistance to Reconciliation in Postgenocide Rwanda*. Madison: University of Wisconsin Press.

Timsar, R.G. 2015. "Oil, Masculinity and Violence: Egbesu Worship in the Niger Delta of Nigeria." In *Subterranean Estates: Life Worlds of Oil and Gas*, edited by Hannah Appel, Arthur Mason, and Michael Watts, 72–90. Ithaca: Cornell University Press.

UN Office of the High Commissioner for Human Rights (OHCHR). 2010. *Democratic Republic of the Congo: Report of the Mapping Exercise Documenting the Most Serious Violations of Human rights and International Humanitarian Law Committed within the Territory of the Democratic Republic of the Congo between March 1993 and June 2003*. United Nations Human Rights Office of the High Commissioner. www.refworld.org/docid/4ca99bc22.html.

Vidal, Claudine. 2001. "Les Commemorations du Genocide au Rwanda: Violence symbolique, mémorisation forcée et histoire officielle." *Les Temps Modernes, 613* (March/April/May): 1–46.

Vltchek, Andre. 2011. "Opaque Waters at Killer Lake." Chinadialogue, August 19. www.chinadialogue.net/article/show/single/en/4472-Opaque-waters-at-Killer-Lake.

Waldorf, Lars. 2006. "Mass Justice for Mass Atrocity: Rethinking Local Justice as Transitional Justice." *Temple Law Review 79* (1): 1–87.

Watts, Michael. 2004. "Resource Curse? Governmentality, Oil, and Power in the Niger Delta, Nigeria." *Geopolitics 9* (1): 50–80.

Weszkalnys, Gisa. 2014. "Anticipating Oil: The Temporal Politics of a Disaster Yet to Come." *The Sociological Review 62* (S1): 211–235.

Wilson, Richard A. 2001. *The Politics of Truth and Reconciliation in South Africa: Legitimizing the Post-Apartheid State*. Cambridge: Cambridge University Press.

Winther, Tanja. 2008. *The Impact of Electricity: Development, Desires and Dilemmas*. New York: Berghahn Books.

# 6  A politics of the public sphere

ENGOs and oil companies in the international climate negotiations, 1987–2001

*Simone Pulver*

Some of the most eye-catching events at the international climate negotiations are the publicity stunts staged by environmental groups. At the 1998 round of the climate negotiations, Friends of the Earth, Greenpeace, and OilWatch arranged a press event in which one person, dressed as the Exxon tiger, led on puppet-strings a suited figure representing the United States. The Exxon tiger was stuffing money into US pockets, and the US was chanting, "Oil good, Exxon good." The goal of the publicity stunt was to portray the US government delegation to the international climate negotiations as a puppet of oil companies. The procession toured around the conference center, gathering a following of delegates and observers, until it stopped in front of the meeting room assigned to conference attendees from business and industry associations. Initially, the industry representatives clustered around the open door and peered out at the spectacle. After a few minutes, they closed the door, and the procession disbanded.

The micro-politics of this episode raise two questions about the interactions between environmental nongovernmental organizations (ENGOs) and oil companies in the context of the interstate deliberations of the United Nations (UN) climate debates. First, what conditions enabled such a display, and second, to what effect? My answer to both questions draws on Habermas's (1989 [1962]) concept of the "political public sphere." According to Habermas, public spheres are inclusive spaces centred on rational deliberation among equals about the common good, with the goal of influencing state authority. I argue that the institutional arrangements of the UN climate negotiations enabled a form of politics associated with a Habermasian public sphere; the arrangements both conferred procedural and discursive advantages on ENGO advocates relative to oil industry representatives and offered a public forum, linked to a rulemaking body, to "name and shame" oil companies for their environmentally destructive practices and their obstructive lobbying. In other words, the rules and procedures governing ENGO and oil company access, participation, and interaction in the UN climate negotiations created the conditions that enabled the display described above. Moreover, the assembled audience in the conference center gave the publicity stunt its purpose. The environmentalist strategy was to make public and visible claims about the undue influence of American oil companies on the negotiating position of the US delegation and to hold Exxon and the American delegation accountable to the international conference community by means of publicity.

My use of the public sphere concept to analyze the interactions between ENGOs and oil companies in the UN climate negotiations unmoors Habermas's theory from the nation state, its original institutional context, and extends it to describe the politics of international governance arenas. In my argument, international deliberative forums do not and cannot function as public spheres in the way envisioned by Habermas, because the publics engaged in international politics do not vote as a way of exercising control over interstate decision making. However, Habermas's theory of the public sphere is not only a theory about public constraint over state power but also a theory of politics (Fraser 1990). Public spheres enable a particular form of politics, namely one that empowers the less materially advantaged, representing broad public constituencies to challenge those pursuing particular interests from positions of structural power (Offe and Wiesenthal 1980; Olson 1971). It is this aspect of Habermas's concept that is relevant to theorizing the ways in which anti-oil activists were able to challenge the political influence of transnational oil corporations in the context of the UN climate negotiations. My research uses the four features of a Habermansian public sphere to interrogate the institutional arrangements of the UN climate negotiations and the form of politics enabled by those institutional arrangements.

My argument about an international politics of the public sphere contributes to the literatures on global governance, transnational advocacy and corporate power in global environmental politics. First, this research presents a conceptual framework for comparatively analyzing international governance arenas based on the constituent elements of a Habermasian public sphere. Habermas identified access, participatory parity, rational deliberation about the common good, and linkage to a rulemaking body as the institutional prerequisites for a political public sphere. These criteria offer four dimensions by which to compare deliberative forums and to evaluate the extent to which they are open or closed to civil society advocacy (Risse 2000). Second, this research re-expands the scope of action of transnational environmental advocacy groups and networks (Wapner, 1995). Rather than focusing only on the specific goal of influencing the outcomes of particular environmental negotiations (Betsill and Corell 2008), I examine the role of transnational activists in creating the international institutions that set the terrain for their advocacy efforts. This is a dynamic conception of transnational activism that describes advocacy groups and networks not only as operating within given international political opportunity structures (Reimann 2006; Shawki 2010) but also as co-creating them. Third, the research contributes a focus on institutional context to the neopluralist perspective on corporate power and its limits in international environmental policy (Falkner 2012).

The research presented in this chapter centers on ENGO challenges of the oil industry in the first 15 years of the international climate change negotiations, from 1987, when the idea of a global climate treaty was first proposed, to 2001, when the final details of the Kyoto Protocol were agreed upon, leading to its eventual entry-into-force in 2005. The research compares the efforts of anti-oil activists and oil company representatives to influence the course of international climate policy. The empirical data analyzed includes transcripts of interviews with 75 ENGO, oil industry, and government participants in the UN climate negotiations; participant observation at four Conferences of the Parties to the UN Framework Convention on Climate Change (UNFCCC) (COPs 4, 6, 6bis, and 7); and archival documents relevant to the climate policies and campaigns of state delegations, ENGOs, oil companies, and business associations involved in the climate debates. The chapter is divided into five sections. Following this introduction, I elaborate on the scholarly contribution of the research, extending the concept of the public sphere to international environmental governance. The next section begins with a brief overview of the UN climate change negotiations and of the activities of anti-oil activists in the context of the international climate debates. I then describe how anti-oil activists leveraged the public sphere aspects of the UN climate negotiations to challenge the political influence of the oil industry. The fourth section describes the contributions of the environmental community to initiating and maintaining the public sphere aspects of the UN climate negotiations, followed by a brief conclusion.

## International governance institutions, transnational advocacy and corporate power in global environmental politics

The international arena is not an obvious forum for transnational environmental advocacy. Dominant theories in international relations consider the international arena the province of warring states (Milner 1991), each concerned with maximizing geopolitical and economic self-interest (Rowlands 2001). Increasing economic globalization over the past 40 years led some scholars to place global capital on par with states in the international system (Strange 1992). Neither version of the international arena—as an arena of interstate competition or as the feeding ground for global capital—suggests an institutional terrain that empowers advocacy movements seeking to constrain state and/or corporate power. However, the increase in numbers and prominence of civil society actors, particularly NGOs, in international forums—for example, the Union of International Associations recorded 48,000 international nongovernmental organizations in 2010, a two-fold increase in 10 years (Union of International Associations, 2011)—belies dominant theory and suggests a reconsideration of the international arena as a site for civil society activism.

Initially, the upsurge of transnational NGOs and networks prompted dramatic expectations that global civil society would "reconstruct, re-imagine, or re-map world politics" (Lipschutz, 1992, 391). Over time, expectations were tempered and research identified the specific factors that determine the emergence and the effectiveness of transnational activism (Betsill and Corell 2008; Keck and Sikkink 1998; Tarrow 2005). International governance institutions, such as multilateral environmental negotiations and treaties, world financial and trade institutions, and international human rights panels and declarations, were identified as playing a crucial role in both emergence and effectiveness. As targets of activism, international governance institutions are catalysts for transnational organizing (Tarrow 2001). As the institutional context for advocacy activities, they constitute the political opportunity structures shaping the influence of transnational activists (Shawki 2010). The particular arrangements of international governance institutions appear not to matter for the catalytic role. Processes inclusive of stakeholder voices are as likely to generate transnational advocacy activity as those closed to non-state actors. Compare for example, union organizing centered on the North American Free Trade Agreement, which welcomed input from labor organizations (R. Evans and Kay 2008), and the labor protests targeting the closed negotiations of the World Trade Organization (Buchanan 2010). Both resulted in enhanced transnational labor organizing (P. Evans 2005; Kay 2005). However, the extent to which governance institutions create opportunities for influence by different types of non-state actors seems to vary significantly. The phenomenon of "forum shopping," when constituencies seek improved outcomes by strategically attempting to shift deliberations from one lawmaking venue to another (Baumgartner and Jones 1993; Pralle 2003), and the boomerang politics of transnational advocacy networks, by which international arenas are used to exert influence in situations when domestic politics are blocked (Keck and Sikkink 1998), point to the potentially enabling or limiting effects of various institutional terrains.

The first contribution of this research is to theorize the aspects of international governance institutions that produce opportunities for influence by advocacy organizations and the network representing broad public constituencies. Habermas's concept of the public sphere offers a starting point. According to Habermas, political public spheres are deliberative arenas that generate public opinion based on inclusive rational debate among equals about the common good in order to influence state action. Public spheres thus embody a particular form of politics; a politics open to all, premised on equality, centered on rational deliberation about the common good, and reliant on public accountability as a means of influence (Habermas 1974 [1964]). Politics practiced in a public sphere stand in stark contrast to a form of politics that eschews public involvement and consent and is conducted through instrumental backroom deals among the powerful few (Vogel 1996; Yergin 1992). Risse (2000), describing international governance arenas, characterizes these two forms of politics as open and closed. He describes closed institutions as limiting access, tending to focus on instrumental bargaining, and giving little consideration to international public goods norms. In contrast, open institutions embody elements of deliberation, argumentation and truth-seeking behavior.

Building on Risse's work, I draw on Habermas to examine the particular institutional arrangements that characterize open versus closed international deliberative arenas. Habermas's theory of the public sphere identifies four key aspects of such arrangements: Accessibility, the degree of parity among participants, the focus of and foundations for deliberations, and the means of influence on rulemaking bodies. For a politics of the public sphere, access has to be inclusive, participants in the public sphere must deliberate as peers, disregarding differences in status, deliberations must focus on the common good and exclude merely private interests, and finally, the goal of deliberation must be to influence on rulemaking. It is worth noting that none of these four criteria for a public sphere is present in its ideal form in any deliberative forum, not in Habermas's original example of eighteenth-century European bourgeois society (Eley 1992; Fraser 1990) nor in the UN climate change negotiations. Nevertheless, Habermas's theory of the public sphere establishes an ideal that serves as a reference point for analyzing the institutional arrangements of international deliberative arenas. Moreover, while each of Habermas's four criteria for a public sphere has been the subject of extensive research in international politics—see Reimann (2006), Smith (2004a) and Raustialia (1997b) on access, Raustiala (1997a), Bäckstrand (2006a) and LePrestre (2014) on participation, Risse (2000) on rational deliberation, and Bartley and Child (2014), Bernstein (2011), Gupta (2010) and Bäckstrand (2006b) on naming, shaming and accountability as a means of influence—the innovation is to integrate them and to theorize the form of politics they engender.

The second contribution of this research is to scholarship on ENGO influence in global environmental politics. As in other international issue areas, transnational environmental activism co-evolved with the increase in multilateral treaty negotiations centered on environmental concerns. Meyer et al. (1997) describe this process as the structuring of a world environmental regime, showing that the rise of international environmental associations is closely linked in time to the initiation of international environmental treaties. In turn, multilateral environmental treaty negotiations became both arenas in which to demand increasing formal recognition for ENGOs in the UN system (Raustiala 1997b) and sites for ENGO campaigns seeking to influence treaty outcomes (Betsill and Corell 2008). With this broader context, recent research has come to focus primarily on the influence of ENGOs on treaty outcomes. Focusing in on specific environmental negotiations, scholarship identifies ENGO access, activities and resources as key determinants of ENGO influence, mediated by issue characteristics and the nature of targets (Arts 1998; Clark, Friedman, and Hochstetler 1998; Corell and Betsill 2001; Vormedal 2008).

However, a narrow focus on direct influence on negotiated outcomes excludes the broader range of strategies by which advocacy groups and networks influence global environmental politics. An expanded definition of influence can be gained by zooming in on influence via the micro-processes of environmental negotiations, what Witter et al. (2015) call moments of influence. Likewise, zooming out to focus on the role of NGOs in creating the institutional context in which they vie for influence also re-expands the potential of ENGO action (Clark 1995; Wapner 1995). In the climate change case, I argue that the environmental community shaped international climate policy outcomes not only through advocacy in the climate negotiations but also by contributing to the creation and maintenance of the institutional terrain on which climate change is debated. In the UN climate negotiations, ENGOs pursued a two-pronged politics of the public sphere. In the short term, environmental groups took advantage of the institutional arrangements of the UN negotiations to "name and shame" oil companies for pursuing particular interests from positions of structural power and challenged them to justify their positions on greenhouse gas (GHG) regulation in terms of widely shared interests. However, this short-term strategy was premised on the existence of a forum like the UN climate negotiations. The second component of a politics of the public sphere is thus the long-term strategy of establishing and maintaining deliberative forums that enable a politics of the public sphere. In the case of climate change, ENGOs played a key role in advocating for international negotiations under the auspices of the UN and actively maintained the publicity of the negotiations as they progressed.

Finally, as a third contribution of this research, a focus on international governance institutions extends scholarship on the limits to corporate power in the international arena. The sources of business influence in global environmental politics are numerous, including structural power based on the economic contribution of business, elite power based on financial contributions to political campaigns, technological power based on business investment in research and development, and discursive power based on economic expertise and advertising and informational campaigns (Clapp and Meckling 2013; Fuchs 2007; Levy and Egan 1998). Theoretical perspectives on business's ability to wield this influence range from political economy approaches that envision transnational capital as determining global politics (Sklair 1994) to pluralist perspectives that identify business as one of many, essentially equivalent interest groups (Hanegraaff 2015). Neopluralists offer a compromise by acknowledging the power of business but also theorizing the limits to business influence (Lindblom 1977). Such limits include competing constituencies (Ronit 2007; Sell and Prakash 2004) and business conflict (Falkner 2008). I argue that the institutional arrangements of international governance arenas are a third potential limit to business influence. First, they can restrict business influence relative to competing constituencies. Shaping the political influence of competing actors is a key mechanism by which institutional arrangements affect substantive outcomes (Chorev 2005). Second, the formal and informal rules of international governance institutions can exacerbate business conflict. Comparing the efforts of anti-oil advocates and oil companies pursuing their interests in the UN climate change negotiations highlights both mechanisms. The structural and elite power of the oil industry was devalued in favor of the discursive power of representing the shared public interest, a claim more convincingly made by anti-oil advocates. Moreover, the formal rules of participation in the UN climate change negotiations exacerbated internal conflicts among oil companies, undermining their international lobbying efforts.

# A politics of the public sphere in the UN climate negotiations

Habermas's public sphere concept provides a framework to analyze how anti-oil groups challenged the structurally and materially powerful oil industry's efforts to influence international climate policy in the context of the UN climate negotiations. Governance arenas with the characteristics of public spheres enable of a form of politics based on open access, participatory parity, and rational deliberation about the common good, with the goal of influencing rulemaking. A politics of the public sphere tends to empower materially disadvantaged groups representing the broad public interest relative to those pursuing private interests based on structural power. This section begins by providing an overview of anti-oil campaigning in the context of the UN climate negotiations. I then analyze the institutional arrangements of the negotiations, examining if and how these arrangements aligned with a public sphere and enhanced the political influence of anti-oil advocates relative to oil industry representatives.

## Challenging the oil industry in the UN climate negotiations

The UN negotiations on climate change are centered on a series of structured deliberations, bringing together national delegations from UN member countries and a range of non-state observers and intergovernmental organizations. From 1990 to 1995, national delegations deliberated under the framework of the Intergovernmental Negotiating Committee (INC), established by UN Resolution #45/212 in 1990. The INC met 12 times between 1990 and 1994. The primary product of the INC deliberations was the 1992 UNFCCC, which was opened for signature at the 1992 UN Conference on Environment and Development, more commonly known as the Rio Earth Summit (Bodansky 2001). Once the UNFCCC entered into force in 1994, the Conference of the Parties (COP) became the primary negotiating body. The COP has met on a more or less annual basis since COP 1 in April of 1995. The UNFCCC also called for the establishment of other negotiating bodies that give advice to the COP. There are two permanent subsidiary bodies: the Subsidiary Body for Scientific and Technological Advice (SBSTA) and the Subsidiary Body for Implementation (SBI). There have also been temporary subsidiary bodies, such as the Ad hoc Group on the Berlin Mandate (AGBM) that drafted the general framework for the Kyoto Protocol (Yamin and Depledge 2004). Finally, there are inter-sessional meetings laying the groundwork for the negotiations at the COPs. While many of the tasks related to formulating international climate policy are completed by subsidiary bodies at inter-sessional meetings or through bilateral dialogue, the COPs serve as the venues at which key climate treaties and declarations are finalized and given approval. During the study period of this research project, the 1992 UNFCCC and the 1997 Kyoto Protocol to the UNFCCC, agreed upon at COP 3, were the two key international climate treaties negotiated. Details related to the implementation of the Kyoto Protocol were finalized at COP 7 in 2001 (Betsill 2005).

Each round of the international climate negotiations has a formal and informal component, created by the hierarchy between official state delegations and non-state observers in the UN system. The formal component of the UN climate negotiations consists of a series of deliberations and/or presentations, structured by a daily agenda, set by the COP leadership and governed by UN rules of procedure. Each COP lasts approximately 2 weeks, consisting of a preparatory segment followed by a high-level ministerial segment (Yamin and Depledge 2004). The primary actors in the formal component of the UN climate negotiations are national delegations, composed of one or more officials, representing their home governments and negotiating on their behalf. During the period of study, state delegations ranged in size from one to over 30 individuals and included representatives from foreign affairs ministries, environmental ministries, and trade and economic ministries. Between 1990 and 2001, the number of national delegates attending the INCs and COPs has ranged from just over 400 delegates attending INC 1, representing 102 countries, to over 2000 national delegates, representing 173 countries, at COP 6 in 2000.[1]

In addition to the formal component of the UN climate negotiations, there is a parallel informal component, embodied by the activities of international organizations, media representatives, and non-state observers to the negotiations (Newell 2000). The halls of the conference centers serving as venues for the INCs and COPs are lined with booths sponsored by a broad range of organizations, peddling their informational fact sheets and reports. Throughout the 2-week negotiating sessions, these non-state observers host briefing sessions, stage side events, generate media coverage, and lobby state delegations. Although national delegations contribute to discussions in the informal arena, its dominant agents are non-state observers. The UN Climate Change Secretariat distinguishes five types of non-state observers organization: Business and Industry NGOs (BINGOs), Environmental NGOs (ENGOs), Indigenous Peoples Organizations (IPOs), Local Governments and Municipal Authorities (LGMAs), and Research and Independent NGOs (RINGOs) (Yamin and Depledge, 2004). As with the numbers of national delegates, there was a dramatic increase in the number of non-state observers attending the negotiations. From INC 1, held in 1991 in Chantilly, Virginia to COP 7, hosted by Morocco in 2001, the number of non-state observers attending the negotiations rose from 149 individuals, representing 71 organizations, to 1327 individuals, representing 194 organizations, with peak attendance of 3663 observers/236 distinct organizations at COP 3 in 1997 in Kyoto, Japan.

Among non-state observers environmental advocates were active participants in the informal component of the UN climate negotiations. Between 1991 and 2001, over 200 unique environmental NGOs attended at least one round of the UN climate negotiations, beginning with 35 ENGOs in attendance at

INC 1 in 1991 and peaking at COP 6 in 2000, with 85 formally registered ENGOs. For ENGOs targeting the oil industry, the UN climate negotiations served as a venue to draw international attention to ongoing, multi-sited local struggles against the polluting practices of oil companies, to expose the corporate lobbying tactics, and to critique the corporate climate policies of leading oil companies. Between 1991 and 2001, anti-oil activists initiated campaigns, published reports, and organized direct action and media events centered on eight distinct but overlapping aspects of the oil–climate change nexus (Table 6.1).[2] These included bringing the oil industry to the fore in debates about climate change, attacking industry lobbying practices, exposing links between the oil industry and scientists skeptical of climate change, using climate change to justify a moratorium against new oil and gas exploration, targeting government subsidies of fossil fuels, framing climate change as an environmental justice issue, and attacking the climate policies of individual oil corporations. Across the eight general areas of anti-oil activism, specific campaign activities at the UN climate negotiations ranged from blocking fossil lobbyists from exiting their limousine when arriving at the conference center (at COP 1 in 1995), to issuing a declaration calling for an end to fossil fuel subsidies (at COP 3 in 1997), to sponsoring students to attend and disrupt the COP 6 negotiations,

*Table 6.1* Anti-oil campaigning in the context of the UN Climate Negotiations (1991–2001)

| Theme of anti-oil activism | Start of activities | Participating ENGOs |
| --- | --- | --- |
| Linking the oil industry and climate change | 1993 | Greenpeace International, Friends of the Earth US, Transnational Resource and Action Center (TRAC, later CorpWatch), Natural Resources Defense Council with Union of Concerned Scientists and US Public Interest Research Group, Greenpeace US |
| Oil industry lobbying | 1995 | Alliance Against Carbon Criminals, Greenpeace International, Greenpeace US |
| IPCC and climate skeptics | 1996 | Greenpeace International |
| No New Frontiers | 1997 | Greenpeace International, Greenpeace UK |
| Eliminating subsidies for fossil fuels | 1997 | OilWatch, Greenpeace US |
| Climate justice | 1998 | TRAC, Project Underground, Rainforest Action Network, Friends of the Earth International, GermanWatch, CorpWatch, Climate Justice Coalition |
| ExxonMobil obstructionism | 1998 | GermanWatch, Campaign ExxonMobil, National Environmental Trust, Friends of the Earth International, People and Planet |
| Shell and BP greenwashing | 1998 | Greenpeace International, SaneBP, Greenpeace Netherlands |

to shareholder campaigns targeted at specific companies. However, the most common campaign activity was the publication of detailed exposés/reports challenging corporate claims.

Leading ENGOs targeting the oil industry at the UN climate negotiations included Greenpeace International and several national Greenpeace offices, Friend of the Earth International, OilWatch and the group of NGOs that formed the Climate Justice Coalition. Most of these groups were members of the Climate Action Network (CAN), a network established in 1989 for NGOs who share a common concern for the problems of climate change and who seek to influence "the design and development of an effective global strategy to reduce greenhouse gas emissions and ensure its implementation at international, national and local levels in the promotion of equity and sustainable development" (Climate Action Network 2012). CAN became the primary organization through which the environmental community organized its international advocacy activities. However, it was never a platform for anti-oil campaigns, because only some of its members endorsed critical campaigns targeting oil companies. Different ENGOs brought different motives to their anti-oil, climate activism. Greenpeace International was the first ENGO to explicitly target fossil fuel companies for their contributions to climate change in the context of the UN climate negotiations, issuing the report "Fossil Fuels in a Changing Climate" in 1993. Reports with a similar theme were subsequently issued by the Transnational Resource and Action Center (TRAC, later CorpWatch), the Natural Resources Defense Council with the Union of Concerned Scientists and the US Public Interest Research Group, and Greenpeace US. The general linkage between the fossil fuel industry and climate change was further refined into two more targeted campaigns. Greenpeace spearheaded the "No New Frontiers" campaign to end oil exploration and extraction at ecological and technological frontiers, while OilWatch, a network of groups engaged in localized struggles against fossil fuel projects, and others sponsored an ongoing campaign to end public subsidies for oil extraction. Early on in the negotiations, Greenpeace also spearheaded a campaign focused explicitly on the UN climate negotiations process, targeting the undue influence of the oil industry on climate science and business lobbying in the negotiations. The oil industry was an early and active participant in the international climate negotiations. Acting through a series of industry associations, representatives from the major multinational oil corporations, including Exxon, Mobil, Texaco, Chevron, British Petroleum, Royal Dutch/Shell, were attendees at the UN negotiations from INC 1 onwards (Pulver 2002). On a separate track and after the negotiation of the Kyoto Protocol, a group of NGOs came together in to form the Climate Justice Coalition, which sponsored an alternative climate justice summit at COP 6 in 2000 and published several reports defining climate justice and linking climate change to local struggles at sites of fossil fuel extraction and production. Finally, as individual oil companies articulated distinct climate policy positions, they became direct targets themselves. Exxon (known as Esso in Europe and then as ExxonMobil after its merger with the Mobil Oil Corporation in 1999) was a focus of anti-oil activism because of its obstructionist activities. BP and Royal Dutch/Shell became targets because of their attempts to greenwash their climate change policies and practices (Pulver 2007).

Anti-oil activism in the UN climate negotiations between 1991 and 2001 had a mixed record of success. Activism had no real impact on oil exploration activities or public subsidies for fossil fuels (BP 2015). Likewise, anti-oil campaigns failed to quell fossil-fuel funded climate skepticism (Dunlap and McCright 2015). However, anti-oil activists did claim several victories. First, in the summer of 1997, two leading international oil corporations, BP and Royal Dutch/Shell, broke ranks with the rest of the oil industry, announcing that they favored a precautionary approach to climate change and that they were leaving the Global Climate Coalition, the most obstructionist industry lobby group in the UN negotiations (Pulver 2007). Later that year, the environmental community celebrated the negotiation of the Kyoto Protocol, which mandated a collective 5.2 percent reduction in GHG emissions by 2008–2012 from 1990 baselines for industrialized country signatories (Betsill 2005). Despite its flaws, the protocol was seen as a victory of environmental over fossil fuel interests. Greenpeace's post-Kyoto press release reasoned that "the fact that a legally binding agreement was reached over the objections of the oil companies shows that industry's grip on governments and stranglehold on the political process is finally loosening" (GP 1997). The industry response to the Kyoto Protocol confirmed the environmentalist victory. The Global Climate Coalition released a press statement that described the Kyoto Protocol as "unilateral economic disarmament. It is a terrible deal and the [US] President should not sign it. If he does, we will campaign, and we will defeat it" (GCC, 1997). The final details of the Kyoto Protocol were agreed upon in 2001, leading to its entry-into-force 4 years later. The year 2001 also marked the disbanding of the Global Climate Coalition, after a flood of defections following BP and Shell (Pulver 2002).

### Leveraging the power of the public sphere

The negotiation of the Kyoto Protocol indicates that fossil fuel interests did not dictate the trajectory of international climate policy, at least initially. I argue that the ability of anti-oil groups to challenge the political influence of the oil industry in the international climate debates was enabled by the institutional arrangements of the negotiations. Habermas's theory of the public sphere identifies four key features by which to evaluate the institutional arrangements of deliberative forums; their accessibility, their degree of parity among participants, their focus of and foundation for deliberation, and their connection to rulemaking bodies. Using this framework to analyze the UN climate negotiations demonstrates why international climate conferences were a favorable site for anti-oil environmental advocacy.

*Access*

The first institutional criterion for a Habermasian public sphere is open access. A public sphere serves as a forum, potentially open to all members of the public, where public opinion is formed to influence state decision-making (Habermas 1974 [1964]). In the context of the UN climate negotiations, the agreements negotiated by state delegates are the form of state authority to be influenced. The public sphere of the UN climate negotiations is created by the exchange of ideas among the non-state observers attending the UN climate conferences. However, unlike Habermas's ideal, access to this public sphere is restricted. First, access is granted through organizations rather than given to individuals directly. And while environmental groups claim to represent their constituencies and oil corporations their shareholders, neither type of organization is democratically accountable to a public (Charnovitz 2006; Cooper and Owen 2007). Second, even organizational access is restricted, by bureaucratic requirements, travel costs and procedural rules. The following analysis examines the rules governing the access of non-state observers to the UN climate negotiations, with an emphasis on the relative access of anti-oil advocacy groups and oil corporations.

Guaranteed access to UN negotiations is a right reserved for state delegates representing UN member countries. Any UN member country may send a delegation of any size to the climate negotiations. In contrast, environmental groups and oil companies, as non-state observers to the negotiations, are granted only conditional access.[3] First, in order to attend the negotiations, they need to be accredited by the UN Climate Change Secretariat, an international organization created to support the UNFCCC. The criteria for observer accreditation, specified in the UNFCCC, Article 7, Paragraph 6, are "being qualified in matters covered by the convention," submitting proof of non-profit, tax-exempt status, and the approval of at least two-thirds of the Parties to the convention (UNFCCC 1994a). For most ENGOs and anti-oil groups, the accreditation criteria were easy to fulfill. However, they were a hindrance to oil industry representatives, particularly the tax-exempt requirement, which stipulates that oil companies cannot represent themselves directly in the UN climate change negotiations but must join a non-profit business association, or BINGO, if they want to attend. BINGO membership can be both an asset and a liability. BINGOs allowed oil companies to distance themselves from obstructive lobbying but also created organizational challenges when there were internal rifts about lobbying strategies among oil companies. Disagreements between American and European oil companies led to the dissolution of the Global Climate Coalition, a leading oppositional BINGO, in part because individual companies could not represent themselves directly but were forced to speak through industry associations. Environmental NGOs did not face the same strictures, being able to both work through CAN and speak as individual organizations (Pulver 2002). The industry community expressed frustration about this indirect form of access and initiated a dialogue on alternative means to harness business and industry input into the negotiations. At INCs 10 and

11, the New Zealand delegation proposed a special Business Consultative Mechanism that would help in "developing a closer, more positive relationship with business interests, in particular those major transnational corporations whose products or activities are significant in terms of greenhouse gas emissions" (UNFCCC, 1994b). Ultimately, the Business Consultative Mechanism was not established, in part due to objections by the environmental community (SBSTA 1996a, 1996b).

Once accredited, non-state observers need to self-fund their attendance at negotiations. The costs of attending the international climate negotiations, including international air travel and hotel accommodations, presented a hurdle to widespread attendance by environmental groups, particular grassroots groups from the Global South (Duwe 2001). Oil company representatives faced no such challenges. Once accredited and on-site at a COP, access can still be restricted, since it is a privilege and not a right. For example, at COP 15 in 2009, non-state observer entry into the conference center was restricted because of the upsurge in numbers of attendees (Fisher 2010; McGregor 2011). Likewise, non-state observers have been ejected from the conference center because of unbecoming conduct. More generally, even if non-state observers are allowed into the conference center, their access to specific deliberations may be limited. Plenary sessions during which lead delegates read declarative statements about their official stance on climate change are always open to non-state observers. However, sessions termed "informal informals," during which bases for agreement are hammered out between lead negotiators of opposing delegations, are closed, limited to a few national delegations.

In summary, the institutional arrangements of the UN climate negotiations offered some public access via membership in non-profit organizations with expertise related to climate change. While not completely open access—for example attendance at the World Social Forum is less restricted (Smith 2004b)—the UN climate negotiations are closer to the open access characteristic of a public sphere than other international forums, such as the North Atlantic Treaty Organization, which excludes non-state observers (Risse 2000). In terms of the relative access of anti-oil activists and oil industry representatives, the data indicate two forms of restricted access. One form favored environmental advocacy groups, by adding the restriction that oil industry representatives could only attend the negotiations as members of non-profit industry associations. This made conflict among oil companies more difficult to manage than conflict among ENGOs. The other form of restricted access benefited oil industry representatives, who were better able to fund their attendance at successive rounds of the international climate negotiations.

*Participatory parity*

Habermas's second prerequisite for a public sphere is some parity among participants. He argues that to enable fair and rational deliberation about the common interest, those engaged in deliberation must set aside social and other inequalities and interact as equals. In other words, participatory parity insures that the force of deliberants' opinions in public debate comes only from the validity of their arguments and not from their social position. In Habermas's theory of the bourgeois public sphere, parity is achieved by "bracketing" differences in social status and leaving them outside the discussion (Habermas 1989 [1962]). As many of Habermas's critics emphasize, this is an aspirational ideal, never achieved in a real social setting (Eley 1992; Fraser 1990). In the UN climate negotiations, oil company representatives did not "bracket" the structural power and material resources of their employers in their interactions with anti-oil advocates. However, the normative context of the UN climate negotiations favored the latter, enabling some parity between the two constituencies. In particular, the personnel attending the UN climate negotiations created an ideological climate more amenable to environmental arguments.

Achieving participatory parity between oil companies and the environmental NGOs that challenge them is a difficult task. In terms of material resources and structural influence, the oil industry overwhelms the environmental community. With annual net incomes in the range of $5–30 billion, oil companies have extensive resources to support their climate advocacy campaigns (http://fortune.com/global500/). In contrast, the average annual campaign budgets of large international ENGOs, like Greenpeace and the World Wildlife Fund, are in the $100–200 million range (Bagely 2014). Likewise, the structural importance of oil to national economies and national security also assures oil companies direct access to high levels in government (Levy and Egan, 1998). Environmental groups have more tenuous relationships with government agencies. They must often trade-off between critical commentary and access to decision-making forums (Craig, Taylor, and Parkes 2004). In the UN climate negotiations, these inequalities were somewhat counter-balanced by the value preferences of the assembled personnel. The personnel of the UN climate negotiations include the state delegates at each round of the negotiations the elected and appointed leaders of each INC and COP, and the staff of the Climate Change Secretariat. In aggregate, these groups held an environmentalist orientation.

Using home ministries as an indicator of value orientation, an analysis of ministerial affiliations of state delegates at successive rounds of the UN climate negotiations documents that environmental representatives outnumbered those from business and economic ministries. At INC 2, in 1991, delegates representing environmental ministries accounted for 36 percent of the 478 attendees, while delegates from economics ministries represented only 11 percent. Representatives from foreign affairs ministries constituted the balance. The pattern became more exaggerated by 1995, when 868 national delegates attended COP 1, with 47 percent from environmental and science ministries and 20 percent from economic ministries. By COP 6 in 2000, attendance had risen to 2148 national delegates, with the majority of delegates representing environmental ministries. This pattern was reproduced in the delegates elected to leadership positions during the COPs. For example, COP Presidents must hold the rank of minister and be responsible for activities in the environmental area (Yamin and Depledge 2004).

A similar environmentalist orientation characterized the staff of the UN Climate Change Secretariat. The secretariat employs approximately 100 people under the leadership of the Executive Secretary, who is appointed by the UN Secretary General in consultation with the COP. The secretariat's primary mandate is to assist the COP and facilitate the work of national delegations. It does so by providing support with logistics, information distribution, and coordination. In addition, the secretariat also coordinates and monitors the attendance of non-state observers at the climate meetings (Yamin and Depledge 2004). As explained above, the formal UN rules governing NGO attendance treat all NGOs equally, and the Climate Change Secretariat conscientiously provides equivalent services to all constituencies. Despite this formal parity, the staff at the Climate Change Secretariat tended to be oriented toward environmental and internationalist concerns. For example, in a workshop on NGO participation in the climate policy process, the Executive Secretary of the Climate Change Secretariat raised the question if NGOs should be required to "declare support for the aims of the Convention, for example, its objective and principles" (UNFCCC 1997, 3). This potential criterion was opposed since it would have prohibited the attendance of those business NGOs opposed to GHG regulation. Staff at the secretariat was aware of the lobbying efforts by industry groups opposed to international GHG reductions. They expressed caution with reference to some industry groups, describing their methods as secretive and at times under-handed and referred to the "mis-information" funded by industry groups prior to COP 3 and COP 6. As a result, business constituencies, particularly those opposed to climate regulation, were sometimes third-tier participants in the climate debates, behind state delegations and environmental NGOs. The BINGO name was a case in point. Business groups didn't like the name, but they were unwilling to make a fuss.

The normative orientation of the personnel of the UN climate negotiations toward environmental concerns made the INCs and COPs attractive venues for anti-oil campaigns. It offered a somewhat more even playing field between the two constituencies, particularly in comparison to many domestic political arenas. Research on oil industry lobbying and anti-oil activism documents the privileged position of oil companies across a range of national contexts (Levy and Egan 1998; McAteer and Pulver 2009; Watts 2005). The UN climate change negotiations suggest a means of enhancing parity based not on the bracketing of status differences (Habermas, 1974 [1964]) nor on an insistence of social equality as a prerequisite for a public sphere (Fraser 1990). Neither is viable in debates over environmental concerns, which often pit well-resourced industry constituencies against less-resourced environmental advocates. Rather, some parity was created because the personnel of the UN climate negotiations was more amendable to environmental arguments.

*Rational deliberation about the common good*

The third constitutive element of a public sphere is rational deliberation about the common good. In a public sphere, arguments must is justified in terms of the public interest and must reflect shared standards for evaluating the validity of claims (Habermas 1974 [1964]). Research that describes climate change as a "wicked problem" would contend that there is no singular conception of a shared public interest in the face of climate change (Levin et al. 2012). However, the institutional terrain of the UN climate negotiations still incorporated an emphasis on the global common interest, however defined. The provision of global collective goods is a central purpose of the UN (Cronin 2002). Moreover, in the UN climate negotiations, a process of rational deliberation was encouraged by the reliance on scientific evaluations of global change. Input from the Intergovernmental Panel on Climate Change (IPCC), in the form of periodic assessments of the state of climate science based on the work of several thousand scientists around the globe, was formally and explicitly used to guide the negotiations (Yamin and Depledge 2004), and the importance of science in the negotiations was enshrined in the texts of both the UNFCCC and the Kyoto Protocol.

Anti-oil groups leveraged both the global collective goods focus and the science-based rationality of the UN climate negotiations in their challenge of the oil industry's structural and elite power. Through a range of campaigns and media events, anti-oil advocates crafted an argument about how oil company activities violated the global common good of "preventing dangerous interference with the climate system," which Article 2 of the UNFCCC established as the principle goal of the negotiations. The core of their argument was a quantitatively based justification against new oil exploration. It was most clearly articulated in "Fossil Fuels and Climate Protection: The Carbon Logic," a report prepared by Greenpeace International in 1997. The report calculated the maximum carbon carrying capacity of the atmosphere needed to avoid dangerous interference with the climate system and compared this number to the carbon in established fossil fuel reserves. The numerical comparison demonstrated that it was "only possible to burn a small fraction of the total oil, coal and gas that has already been discovered" (Hare 1997, i) without severely disrupting the climate system. The carbon logic argument was followed by the "no new frontiers" demand that oil companies avoid the expansion of oil extraction into new and pristine geographies and into new and untested technological arenas. Other campaign demands included a moratorium on all drilling and an end to subsidies for fossil fuel projects. These demanded changes in oil industry practices were framed as necessary to making the industry's operations consistent with global climate protection.

The demands of anti-oil activists were in line with the science-based environmental crisis frame articulated by the wider network of environmental groups attending the UN climate negotiations. The ENGO community argued that climate change was producing an environmental crisis and used the scientific findings of the IPCC to bolster their arguments. Elements of the environmental crisis included the collapse of ecological systems and services, rise in the occurrence and severity of droughts, hurricanes and floods, sea level rise and the devastation of low-lying and small island states, extinction of species, and climatic instability that would disrupt precipitation and agricultural patterns. A survey of articles in *ECO*, the daily newsletter published by CAN at each round of negotiations, underscores this framing. The majority of *ECO*s published between 1991 and 2001 included at least one front-page article focusing on an environmental disruption or disaster. In making its arguments, one of CAN's core principles was to use only peer-reviewed science as a basis for their demands (Betsill 2000). At times, ENGO advocates felt that the IPCC assessments were not strong enough in their description of the potential threat of climate change (Leggett 1990). However, the IPCC was always an ally.

The efforts of anti-oil ENGOs did not go unchallenged. Business interests developed a counter frame emphasizing the economic costs of climate change. From 1989 onwards, conservative business groups released over 20 reports projecting the high costs resulting from binding GHG emissions reductions. For example, in April of 1992, the Global Climate Coalition (GCC) sent a letter to White House urging the Bush administration to resist specific commitments on climate change, predicting $90 billion in annual costs and the loss of 600,000 jobs. This theme was picked up again and again in a 1994 GCC report on the high cost of emissions reductions, in a 1996 symposium organized by the International Petroleum Industry Environmental Conservation Association on "Critical Issues in the Economics of Climate Change," and in a 1998 study commissioned by the American Petroleum Institute. Oil-funded industry groups also sought to challenge IPCC science, sponsoring and promoting the research of a small group of scientists skeptical of the consensus emerging through the IPCC (McCright and Dunlap 2003) and funded *ad hominem* attacks on the IPCC process and leadership (Gelbspan 1997). Both business strategies had limitations. Reports regarding economic costs were often framed in terms of a single nation. While this resonated with broad national concerns, it did not align with the global mandate of the UN. An American Petroleum Institute memo captures the pressure that the UN context placed on industry lobbying groups. The memo contended that,

> from the political viewpoint, it is difficult for the United States to oppose the [climate] treaty solely on economic grounds, valid as the economic issues are. It makes it too easy for others to portray the United States as putting preservation of its own lifestyle above the greater concerns of mankind.

> (API 1998)

The attacks on the IPCC also backfired in the UN context, because they put the oil industry and conservative business lobby at odds with the science-based environmental stewardship embedded in the climate negotiations.

Public spheres emphasize rational deliberation about the common good. In the UN climate negotiations, the common interest referred to preventing dangerous interference with the global climate system and rational deliberation was based on a scientific understanding of climate change established by the IPCC. Both features of the climate negotiations benefited anti-oil activists. Many environmental problems are commons problems, with concentrated benefits and distributed costs (Hardin 1968). As a result, it is easier for environmentalists to argue that they represent the common interest. Likewise, although scientific research is inherently political (Beck 2011; Litfin 1994), the standard business strategy of manufacturing scientific uncertainty (Freudenburg, Gramling, and Davidson 2008) indicates that science-based argumentation tends to benefit environmental constituencies.

*Influence on rulemaking*

Habermas's final criterion for a political public sphere is that opinion formed should influence state rulemaking. In the original democratic nation state context of the public sphere, the means of influence was the right to vote in public elections (Habermas 1974 [1964]). This form of influence is absent in international governance arenas, which confer no voting rights on a global citizenry or on the non-state observers that claim to represent them. In their stead, influence on rulemaking in the UN climate negotiations was sought through two strategies. The first, in direct opposition to a politics of the public sphere, was through back-channel collaborations with like-minded delegations, a strategy pursued by both oil companies and anti-oil advocates. The second means of influence, pursued only by anti-oil groups, centered on "naming and shaming," an established NGO strategy, aimed at both governments (Hafner-Burton 2008) and corporations (Bartley and Child 2014; Spar 1998). In the forum of the UN negotiations, anti-oil advocates named and shamed oil companies for their contributions to global climate change, for the environmental and social consequences of oil extraction and for their obstructionist lobbying, and national governments for their collusion with oil companies. Despite concerns that it is "cheap talk," without effect, statistical evidence from the human rights arena demonstrates that shaming in international arenas reduces violations of political rights by governments (Hafner-Burton 2008). Likewise, research on reputational influences on corporate behavior suggests some corporate concern about negative publicity (Ambec and Lanoie 2008). This means of influence seems mostly likely to be effective in what Fligstein and McAdam (2011) call unorganized social space or arenas with no "stable definitions of the situation" or "sets of 'rules'" that routinize relations between groups making claims, vying for advantage, and attempting to establish control over the social space. And while international environmental deliberations have a long history (Raustiala 1997b), widespread corporate participation dates back only to the early 1990s (Schmidheiny 1992), the beginning of the study period of this research.

The most direct channel of influence for non-state observers during the UN climate negotiations eschewed public sphere strategies in favor of working through like-minded delegations. Close relations with national delegations provided opportunities to submit text to be considered in the negotiations, to be guaranteed a receptive audience for concerns, and to gain intelligence on closed-door sessions. Early on in the process, environmental activists benefited from a close relationship with the Alliance of Small Island States (AOSIS), the group of states most threatened by climate change, and the states in the European Union. For example, the Foundation for International Environmental Law and Development (FIELD), a UK-based environmental NGO, was integrally involved in the formation of AOSIS and in developing the AOSIS negotiating strategy at successive rounds of the negotiations. CAN, Greenpeace, and Friends of the Earth also nurtured close ties to the European Union delegation (Newell 2000). In a parallel effort, the conservative business lobby collaborated closely with the Organization of Petroleum Exporting Countries (OPEC), who had the most to lose from an international climate treaty, and other states like the US and Australia that were generally opposed to the international regulation of GHG emissions (Oberthuer and Ott 1999). One of the strongest links between the oil industry and OPEC in the UN climate negotiations was through the Climate Council, a business NGO based in Washington, DC. The Climate Council's main representative was Donald Pearlman, an international lawyer with the Washington firm Patton, Boggs, and Blow. In the international negotiations, Pearlman acted as an advisor to and lobbyist for the oil producing states in the Middle East (*Der Spiegel* 1995).

However, in addition to these back-channel influence strategies, anti-oil activists also embraced publicity as a means of influence, in line with a politics of the public sphere. In pursuing publicity strategies, anti-oil groups stood in direct contrast to oil industry groups, who sought a low public profile, explicitly describing themselves as observers to the negotiations and not participants. The first 10 years of the UN climate negotiations offer numerous examples of anti-oil activists spotlighting what they considered nefarious practices by oil industry lobby groups and oil companies themselves. For example, they called attention to the undue influence of the Global Climate Council and the Global Climate Coalition, by blocking the limousines bringing representatives of the two BINGOs to the conference center and through an exposé report linking the obstructionist position statements of various BINGOs to their oil company members. Other actions targeted fossil fuel funding of the research and speaking engagements of scientists skeptical of climate change ("The Scourge of the Skeptics: Industry Attacks on the IPCC Second Assessment Report") and the political campaign contributions of US oil companies ("Oiling the Machine"). This form of publicity was regularized as a standard part of the environmental community's contribution to each round of the UN climate negotiations through CAN's "Fossil of the Day" award. The goal of the award was to draw attention to both state and non-state actors that CAN deemed particularly egregious in their attempts to hinder progress in the climate negotiations. The award was presented each evening at a central location in a COP conference center. Daily awards were cumulative, and at the end of the COP an overall "winner" was publicly announced. ENGOs also sought to leverage the influence of a broader public on state climate negotiations, by organizing public protests on the weekend between the first and second weeks of the interstate negotiations. These events were intended to draw public attention beyond the confines of the conference center to the negotiations and to chastise state delegates for the slow progress of the negotiations. The COP 6 round of negotiations in The Hague in 2000 marked the first such protest event. Thousands of activists joined together to build a symbolic dyke around the COP conference center. In 2001, activists built a "climate ark" that was the centerpiece of a protest march in Bonn.

A second thematic focus of ENGO naming and shaming strategies was to bring to the INC and COP conference centers the on-the-ground realities of oil extraction and the continued carbon economy. The moral force of demands to end oil extraction was amplified by tying the activities of oil companies not only to future climate change but also to present ecological and social disruption. Anti-oil activists made this connection through multiple channels. For example, at COP 4 in 1998, Friends of the Earth, OilWatch, and GermanWatch co-organized an event on Tuesday, November 10, centered on Ken Saro-Wiwa, an anti-oil activist from Ogoniland in the Niger Delta, who had been executed by the Nigerian government, exactly 3 years prior. At COP 6 in November of 2000, representatives from a coalition of climate justice NGOs presented a Shell representative with a polluted air sample from Shell chemical refinery in Norco, Louisiana.

In the UN climate negotiations, anti-oil advocates seized on opportunities for influence via publicity by naming and shaming the perpetrators of instrumental lobbying and bargaining and by bringing into the public eye the tangible consequences of fossil fuel extraction and climate change. However, they also recognized the limits of this strategy and pursued back-channel means of influence by working with like-minded state delegations. This duality of strategies points to a tension among the constituent elements of a public sphere. Forums that emphasize access and participatory parity are likely to be farther removed from rulemaking institutions (Risse 2000). Compare for example the World Social Forum, which emphasizes open access and parity but has no ties to any rulemaking body (Smith, 2004b), and the International Labor Organization, where employers and labor are on par with governments in terms of rulemaking, but attendance is restricted to two national delegates, one representing employers and the other labor (P. Evans 2005). The UN climate negotiations reflect a hybrid model, with the informal public sphere of non-state observer activity running concurrent to the formal interstate negotiations, although with only informal mechanisms of influence connecting the two.

## Creating and maintaining public spheres

As described above, a politics of the public sphere entails advocacy activity in particular institutional settings that allow the less materially advantaged, representing broad public constituencies to challenge those pursuing specialized interests from positions of structural power. However, a politics of the public sphere also involves engaging in the longer-term project of creating and maintaining deliberative arenas that function as public spheres. Environmental groups did both in the UN climate negotiations. In addition to taking advantage of the public sphere aspects of the UN climate negotiations to challenge the structural and elite power of the oil industry and its influence on international climate policy, the environmental community also played a key role in advocating for a UN negotiations process centered on climate change in the first place. An epistemic community of atmospheric scientists and activists, based in government agencies, environmental NGOs, and international organizations, first established climate change as a global concern and then called for an international treaty to limit GHG emissions. Once the negotiations commenced, a wider array of environmental groups became involved, who both benefited from the public sphere aspects of the UN climate negotiations and campaigned to maintain them.

### ENGO contributions to initiating the UN climate negotiations

Environmental groups played a pivotal role in the genesis of the international climate negotiations. Many early climate scientists were based in ENGOs, and these individuals pushed for both a science and policy response to the climate problem. The roots of climate change as an international environmental policy issue are in the international scientific community. Climate science first became a significant research topic in the International Geophysical Year in 1957/1958. By the late 1960s, scientists were documenting that atmospheric carbon dioxide concentrations had increased by 10 percent over pre-industrial levels. In 1970, the UN Secretary General mentioned "a catastrophic warming effect" in a speech to the General Assembly. The potential of human-induced global warming was also included in the conclusion of the Stockholm Report issued at the 1972 UN Conference on Environment and Development. However, the 1979 World Climate Conference, jointly organized by the United Nations Environment Program (UNEP) and the World Meteorological Organization (WMO), marked the moment when global warming first gained widespread attention. The next milestones were three scientific conferences held in Villach, Austria (October 1985 and October 1987) and Bellagio, Italy (November 1987). The 1985 and 1987 Villach and Bellagio conferences were the first time that a substantial number of the world's climate scientists agreed that global warming was a serious possibility, estimated the rate of warming ($0.3°C/$decade), estimated contributions of various GHGs to warming, and, most audaciously, suggested negotiating an international treaty to address the problem (Bruce, 2001).

The scientific staff of environmental research and advocacy organizations, including Michael Oppenheimer from Environmental Defense Fund, Gordon Goodman of the Stockholm Environmental Institute (formerly the Beijer Institute), and George Woodwell of the Woods Hole Research Center, played important roles during these early events. They helped to coordinate the

1985 and 1987 Villach and Bellagio workshops (Agrawala 1999). They were also key players in subsequent conferences that marked the shift of the global warming issue from primarily scientific arenas to international political forums. For example, the 1998 Toronto Conference on "The Changing Atmosphere: Implications for Global Security" was largely a product of ENGO organizing. At the conference, members of the NGO community such as Richard Ayres of the Natural Resources Defense Council, Stewart Boyle and Julia Langer of Friends of the Earth, Irving Mintzer and Rafe Pomerance of the World Resources Institute, John Topping of the Climate Institute, and Carol Werner of the Environment and Energy Study Institute, worked as equal partners with scientific participants (Betsill, 2000). The Toronto conference was followed in quick succession by climate-focused conferences in Washington, DC, The Hague, and New Delhi. By 1988, the UN General Assembly first addressed the issue of global warming (resolution #43/53), in response to a proposal by the government of Malta (United Nations Climate Change Secretariat 2003). In 1989, it passed resolution #44/207 calling for the negotiation of a framework convention on climate change (Betsill 2005; Bodansky 2001). Concurrently, the governing boards of UNEP and WMO established the IPCC, which produced its first Scientific Assessment Report in time for the Second World Climate Conference held in Geneva in November of 1990. A month later, the UN General Assembly voted to initiate an international negotiations process on climate change (resolution #45/212) (Bodansky 2001).

The history of ENGO involvement in the emergence of climate change as a scientific and political issue reveals their pivotal contribution to the UN resolution to initiate a climate negotiations process. At the time, ENGOs involved did not have the explicit motive of establishing the UN climate negotiations as a public sphere. However, they did recognize the opportunities for ENGOs in UN environmental negotiations, opportunities that were built on decades of ENGO activism (Willetts 1996). As such their activities contributed to creating the institutional terrain that later allowed them to challenge the fossil fuel industry and its influence on global climate policy.

### Maintaining the public sphere aspects of the UN climate negotiations

Given their contributions to establishing an international climate negotiations process, the environmental community was vigilant about maintaining the public sphere aspects of the UN climate debates. While the international climate negotiations may have been founded on the principle of publicly deliberating the common interest, this was often only an aspirational ideal. In practice, interstate bargaining regarding material interests, completely separated from the informal public sphere, often become the primary mode of decision making. As the stakes in the interstate deliberations intensified, non-state actors (and at times even state actors) were excluded from observing discussions (Yamin and Depledge 2004). Seperately, efforts were made to remove discussion about climate change from the public arena by casting it as an economic issue to be solved by markets, as a topic for expert advice rather than public deliberation, or as a domestic issue and not under the purview of international negotiations (Fraser 1997). Environmental groups counteracted each of these threats to the public sphere aspects of the UN negotiations through repeated efforts to bring discussions happening behind physically and metaphorically closed doors into public settings.

The experience and activities of climate justice NGOs exemplify these efforts. While the environmental community as whole benefited from the public sphere aspects of the UN climate negotiations in relation to oil industry representatives, there were divisions among ENGOs that shaped the respective access, parity, definitions of the shared interest, and influence strategies of individual groups. From the initiation of the UN climate negotiations, the climate justice NGOs were often in disagreement with more mainstream environmental groups (Alcock 2008). The issue of carbon emissions trading exacerbated these tensions. After the negotiation the Kyoto Protocol in 1997, the focus of deliberation in the UN climate negotiations shifted from debating the appropriate response to climate change to working out the details of the emissions trading mechanisms by which countries with Kyoto targets could meet their obligations (Betsill 2005). Several oil companies piloted emissions trading systems in collaboration with Environmental Defense, earning the organizations insider status in the negotiations (Victor and House 2006). The focus on emissions trading caused the negotiations to become more technical, with expertise-based interventions replacing open access deliberation about the shared common good.

Many climate advocacy groups rejected the carbon markets approach to climate change, arguing that it empowered, rewarded, and privileged the very same business actors at the heart of the fossil fuel energy system. The informal coalition of climate justice NGOs, including groups like OilWatch, CorpWatch, the Third World Network, the Rainforest Action Network, and grassroots groups affiliated with Friends of the Earth and People and Planet, was most vocal in advocating this critical perspective. They contended that the negotiations process had become tainted, and they positioned themselves in opposition to the formal negotiations process (CorpWatch 2000). In order to rescue the public sphere aspects of the negotiations, they organized parallel conferences, linked to the formal negotiations, but at venues outside the official COP conference center. Anti-oil and other ENGOs first pursued this tactic in 2000, when they organized a Climate Justice Summit at the COP 6 negotiations. The impetus for this summit was disenchantment with the formal UN climate negotiations process centered on emissions trading and an attempt to create an alternative venue allowing for a social justice focused discussion about climate change. The purpose of the Climate Justice Summit was to give voice to an environmental message that challenged what they considered the private bargaining between states, oil companies and their ENGO partners dominating the COP process. By holding the event outside the COP conference center, they also assured access to all interested participants.

A politics of the public sphere extends beyond seeking out forums whose public sphere aspects benefit constituencies representing the broad public interest. It also involves creating and protecting such institutional terrains. The environmental NGO community played a key role in advocating for international negotiations centered on climate change. They thus helped to create the forum they then used to challenge state and corporate interests in the fossil fuel economy. Moreover, as the UN climate negotiations shifted focus from public discussions of respective responsibilities for reducing GHG emissions to technical discussion about carbon market mechanisms, the community of climate justice NGOs sought to refocus the climate deliberations on the problems of the fossil fuel extraction, the need for more stringent GHG reduction targets, and the social inequities of climate change, bringing the issue of climate change back into the public sphere.

## Conclusion

The dynamics of international environmental negotiations are manifold. The outcomes of treaty negotiations reflect primarily the competing interests of states, as well as the activities of scientific communities, environmental advocates, business actors, and other constituencies. The goal of this chapter was not to attribute the outcomes of the Kyoto Protocol to the activities of the international community of climate advocacy NGOs. Rather, I examined how the institutional arrangements of the UN climate negotiations enabled and hindered the competing constituencies of anti-oil activists and transnational oil company representatives in advocating for their preferred international response to climate change. In the context of the UN climate negotiations, representatives of major oil corporations, arguably the most powerful sectoral interest group in the world, and the BINGOs representing them, were literally and metaphorically herded into a small space from which they defended their rights to continue drilling and refining oil. I argue that this outcome was made possible because the institutional arrangements of UN climate negotiations, which enabled a politics of the public sphere. They provided a relatively open forum in which the value orientations of most of the attendees and organizers created some parity of participation between ENGO representatives and oil companies. Deliberations in the negotiations were science-based and focused on the shared global interest of protecting the planet's climate system. And finally, the negotiations provided anti-oil activists with an audience of state delegates for their naming and shaming of the extractive practices of the oil industry and its obstructionist lobbying. As such the UN climate negotiations offer an empirical example of the ways by which ENGOs were able to leverage the power of the public sphere against a structurally and materially powerful political opponent seeking to undermine international GHG regulation.

This chapter described dual public sphere strategies. First, where a politics of the public sphere is possible in international arenas, advocacy organizations can use the power of publicity to influence decision-making, benefiting from institutional arrangements that legitimate those representing the common interest. Second, advocacy organizations and networks can engage in the project of creating institutional venues that allow for a politics of the public sphere, through long-term institution building or through short-term protest and publicity strategies that shift debate from arenas of private bargaining to arenas where deliberation is subject to public scrutiny. The contributions of Habermas's theory of the public sphere to scholarship on global environmental governance are three-fold. First, the particular form of politics envisioned by his theory of the public sphere identifies four institutional criteria to comparatively assess the institutional arrangements of deliberative governance forums. Second, the portfolio of public sphere strategies available to activists expands the potential scope of influence of environmental advocacy groups in international environmental negotiations. Influence is not limited to shaping the particular text of a treaty but extends to shaping the institutional arrangements that provide the political opportunity structures for advocacy. Finally, Habermas's theory of the public sphere sheds light on how institutional arrangements can limit business influence in global environmental politics.

This research compared the efforts of two competing constituencies differently positioned with respect to a politics of the public sphere within a single deliberative arena. However, Habermas's four criteria for a political public sphere can also serve as a framework for comparing across international governance institutions, across levels of governance, and across the evolution of governance arenas. Each of these theoretical and empirical extensions will help to further refine Habermas's criteria for public spheres, the interrelations among them, and their relevance to contemporary politics.

## Notes

1   Data on delegate, country, individual and organization participation in the UN climate change negotiations were compiled by the author from the Lists of Participants published by the UN Climate Change Secretariat for each round of the climate negotiations.
2   Data on anti-oil activism were compiled from a variety of sources, including mentions of campaign activities in the Climate Action Network's *ECO* newsletters, the UN Climate Change Secretariat archives for each round of the climate negotiations, participant observation at COPs 4, 6, 6-bis and 7, and interviews with ENGOs targeting the oil industry, producing a database of 19 distinct activities organized at/or focusing on the UN climate negotiations and an additional 30 activities targeting major transnational oil corporations for which the UN climate negotiations served as a context. The eight themes of anti-oil activism were inductively derived from the database.
3   A few national delegations, such as Australia and Brazil, consistently include representatives from the private sector and environmental community. However, their freedom of expression is curtailed by the negotiating position of the country they represent.

# References

Agrawala, Shardul. 1999. "Early Science-Policy Interactions in Climate Change: Lessons from the Advisory Group on Greenhouse Gases." *Global Environmental Change* 9 (2): 157–169.

Alcock, Frank. 2008. "Conflicts and Coalitions within and across the ENGO Community." *Global Environmental Politics* 8 (4): 66–91.

Ambec, Stefan, and Paul Lanoie. 2008. "Does It Pay to Be Green? A Systematic Overview." *Academy of Management Perspectives* 22 (4): 45–62.

American Petroleum Institute (API). 1998. *Global Science Communications: Action Plan.* Internal memorandum. Washington, DC: American Petroleum Institute.

Arts, Bas. 1998. *The Political Influence of Global NGOs: Case Studies on the Climate and Biodiversity Conventions.* Utrecht: International Books.

Bäckstrand, Karin. 2006a. "Democratizing Global Environmental Governance? Stakeholder Democracy after the World Summit on Sustainable Development." *European Journal of International Relations* 12 (4): 467–498. doi:10.1177/1354066106069321.

Bäckstrand, Karin. 2006b. "Multi-Stakeholder Partnerships for Sustainable Development: Rethinking Legitimacy, Accountability and Effectiveness." *European Environment* 16 (5): 290–306.

Bagely, Katherine. 2014. Infographic: A Fieldguide to the US Environmental Movement. Website. Inside Climate News, April 4. http://insideclimatenews.org/news/20140407/infographic-field-guide-us-environmental-movement.

Bartley, Tim, and Curtis Child. 2014. "Shaming the Corporation: The Social Production of Targets and the Anti-Sweatshop Movement." *American Sociological Review* 79 (4): 653–679.

Baumgartner, Frank R., and Bryan D. Jones. 1993. *Agendas and Instability in American Politics.* Chicago: University of Chicago Press.

Beck, Silke. 2011. "Moving Beyond the Linear Model of Expertise? IPCC and the Test of Adaptation." *Regional Environmental Change* 11 (2): 297–306.

Bernstein, Steven. 2011. "Legitimacy in Intergovernmental and Non-State Global Governance." *Review of International Political Economy* 18 (1): 17–51.

Betsill, Michelle. 2000. "Greens in the Greenhouse: Environmental NGOs, Norms and the Politics of Global Climate Change." PhD Thesis, University of Colorado Press.

Betsill, Michelle. 2005. "Global Climate Change Policy: Making Progress or Spinning Wheels?" In *The Global Environment: Institutions, Law and Policy*, edited by Regina S. Axelrod, David Leonard Downie, and Norman J. Vig, 103–124. Washington, DC: CG Press.

Betsill, Michelle, and Elisabeth Corell, eds. 2008. *NGO Diplomacy: The Influence of Nongovernmental Organizations in International Environmental Negotiations.* Cambridge, MA: MIT Press.

Bodansky, Daniel. 2001. "The History of the Global Climate Change Regime." In *International Relations and Global Climate Change*, edited by Urs Luterbacher and Detlef F. Sprinz, 23–40. Cambridge, MA: MIT Press.

BP. 2015. *BP Statistical Review of World Energy June 2015.* www.bp.com/content/dam/bp/pdf/energy-economics/statistical-review-2015/bp-statistical-review-of-world-energy-2015-full-report.pdf.

Bruce, Jim. 2001. "Intergovernmental Panel on Climate Change and the Role of Science in Policy." *Isuma: Canadian Journal of Policy Research* 2 (4): 11–15.

Buchanan, Ruth M. 2010. "Protesting the WTO in Seattle: Transnational Citizen Action, International Law and the Event." In *Events: The Force of International Law*, edited by Fleur Johns, Richard John Joyce, and Sundhya Pahuja, 221–233. Abingdon, UK: Routledge.

Charnovitz, Steve. 2006. "Nongovernmental Organizations and International Law." *The American Journal of International Law 100* (2): 348–372.

Chorev, Nitsan. 2005. "The Institutional Project of Neo-Liberal Globalism: The Case of the WTO." *Theory and Society 34* (3): 317–355.

Clapp, Jennifer, and Jonas Meckling. 2013. "Business as a Global Actor." In *The Handbook of Global Climate and Environment Policy*, edited by Robert Faulkner, 286–303. West Sussex, UK: John Wiley & Sons.

Clark, Ann Marie. 1995. "Non-Governmental Organizations and the Influence on International Society." *Journal of International Affairs 48* (2): 507-525.

Clark, Ann Marie, Elisabeth J. Friedman, and Kathryn Hochstetler. (1998). "The Sovereign Limits of Global Civil Society: A Comparison of NGO Participation in UN World Conferences on the Environment, Human Rights, and Women." *World Politics 51* (1): 1–35.

Climate Action Network (CAN). 2012. CAN Charter. Ottawa: CAN. www.climate network.org/about/can-charter.

Cooper, Stuart M., and David L. Owen. 2007. "Corporate Social Reporting and Stakeholder Accountability: The Missing Link." *Accounting, Organizations and Society 32* (7–8): 649–667.

Corell, Elisabeth, and Betsill, Michelle. 2001. "A Comparative Look at NGO Influence in International Negotiations: Desertification and Climate Change." *Global Environmental Politics 4* (1): 86–107.

CorpWatch. 2000. "Alternative Summit Opens with Call for Climate Justice." CorpWatch Press Release, November 19. The Hague, Netherlands: CorpWatch.

Craig, Gary, Marilyn Taylor, and Tessa Parkes. 2004. "Protest or Partnership? The Voluntary and Community Sectors in the Policy Process." *Social Policy & Administration 38* (3): 221–239. doi:10.1111/j.1467-9515.2004.00387.x.

Cronin, Bruce. 2002. "The Two Faces of the United Nations: The Tension Between Intergovernmentalism and Transnationalism." *Global Governance 8* (1): 53–71.

Der Spiegel. 1995. "Hohepriester im Kohlenstoff-Klub." April 3, 36–38.

Dunlap, Riley E., and Aaron M. McCright. 2015. "Challenging Climate Change: The Denial Countermovement." In *Sociological Perspectives on Climate Change*, edited by Riley E. Dunlap and Robert Brulle, 300–332. Oxford: Oxford University Press.

Duwe, Matthias. 2001. "The Climate Action Network: A Glance behind the Curtain of a Transnational NGO Network." *Review of European Community and International Environmental Law 10* (2): 177–189.

Eley, Geoff. 1992. "Nations, Public, and Political Cultures: Placing Habermas in the Nineteenth Century." In *Habermas and the Public Sphere*, edited by Craig Calhoun, 289–339. Cambridge, MA: MIT Press.

Evans, Peter. 2005. "Counter-Hegemonic Globalizations: Transnational Social Movements in the Contemporary Global Political Economy." In *Handbook of Political Sociology*, edited by Thomas Janoski, Alexander M. Hicks, Mildred A. Schwartz, 655–670. Cambridge: Cambridge University Press.

Evans, Rhonda, and Tamara Kay. 2008. "How Environmentalists 'Greened' Trade Policy: Strategic Action and the Architecture of Field Overlap." *American Sociological Review 73* (6): 970–991.

Falkner, Robert. 2008. *Business Power and Conflict in International Environmental Politics.* New York: Palgrave Macmillan.

Falkner, Robert. 2012. "Business Power, Business Conflict: A Neo-Pluralist Perspective on International Environmental Politics." In *Handbook of Global Environmental Politics*, edited by Peter Dauvergne, 319–329. Cheltenham, UK: Edward Elgar Publishing.

Fisher, Dana R. 2010. "COP 15 in Copenhagen: How the Merging of Movements Left Civil Society Out in the Cold." *Global Environmental Politics 10* (2): 11–17.

Fligstein, Nigel, and Doug McAdam. 2011. "Toward a General Theory of Strategic Action Fields." *Sociological Theory 29* (1): 1–26. doi:10.1111/j.1467-9558.2010.01385.x.

Fraser, Nancy. 1990. "Rethinking the Public Sphere: A Contribution to the Critique of Actually Existing Democracy." *Social Text 25/26*: 56–80.

Fraser, Nancy. 1997. *Justice Interruptus: Critical Reflections on the "Postsocialist" Condition.* New York: Routledge.

Freudenburg, William R., Robert Gramling, and Debra J. Davidson. 2008. "Scientific Certainty Argumentation Methods (SCAMs): Science and the Politics of Doubt." *Sociological Inquiry 78* (1): 2–38.

Fuchs, Doris. 2007. *Business Power in Global Governance.* Boulder: Lynne Rienner.

Gelbspan, Ross. 1997. *The Heat is On: The High Stakes Battle over the Earth's Threatened Climate.* Reading, MA: Addison-Wesley.

Global Climate Coalition (GCC). 1997. "Climate Agreement Called Economic Disarmament." In *Rising Voices against Global Warming*, edited by Azza Taalab, 83–84. Frankfurt: Informationszentrale der Elektrizitaetswirtschaft.

Greenpeace (GP). 1997. "Climate Agreement Endangers the Climate." In *Rising Voices against Global Warming*, edited by Azza Taalab, 129. Frankfurt: Informationszentrale der Elektrizitaetswirtschaft.

Gupta, Aarti. 2010. "Transparency in Global Environmental Governance: A Coming of Age?" *Global Environmental Politics 10* (3): 1–9.

Habermas, Jürgen. 1974 [1964]. "The Public Sphere: An Encyclopedia Article." *New German Critique 3* (Autumn): 49–55.

Habermas, Jürgen. 1989 [1962]. *The Structural Transformation of the Public Sphere: An Inquiry into a Category of Bourgeois Society*, translated by Thomas Burger with the assistance of Frederick Lawrence. Cambridge, MA: MIT Press.

Hafner-Burton, Emile M. 2008. "Sticks and Stones: Naming and Shaming the Human Rights Enforcement Problem." *International Organization 62* (4): 689–716. doi:10.1017/S0020818308080247.

Hanegraaff, Marcel. 2015. "Interest Groups at Transnational Negotiation Conferences: Goals, Strategies, Interactions, and Influence." *Global Governance: A Review of Multilateralism and International Organizations 21* (4): 599–620. doi:10.5555/1075-2846-21.4.599.

Hardin, Garrett. 1968. "Tragedy of the Commons." *Science 162* (3859): 1243–1248.

Hare, Bill. 1997. *Fossil Fuels and Climate Protection: The Carbon Logic.* Amsterdam: Greenpeace International.

Kay, Tamara. 2005. "Labor Transnationalism and Global Governance: The Impact of NAFTA on Transnational Labor Relationships in North America." *American Journal of Sociology 111* (3): 715–756.

Keck, Margaret E., and Kathryn Sikkink. 1998. *Activists Beyond Borders: Transnational Advocacy Networks in International Politics.* Ithaca: Cornell University Press.

Leggett, Jeremy, ed. 1990. *Climate Change: The Greenpeace Report*. Oxford: Oxford University Press.

LePrestre, Phillipe. 2014. "Participation." In *Essential Concepts of Global Environmental Governance*, edited by Jean-Frédéric Morin and Amandine Orsini, 144–145. Oxford and New York: Routledge/Earthscan.

Levin, Kelly, Benjamin Cashore, Steven Bernstein, and Graeme Auld. 2012. "Overcoming the Tragedy of Super Wicked Problems: Constraining Our Future Selves to Ameliorate Global Climate Change." *Policy Sciences 45* (2): 123–152. doi:10.1007/s11077-012-9151-0.

Levy, David L., and Daniel Egan. 1998. "Capital Contests: National and Transnational Channels of Corporate Influence on the Climate Change Negotiations." *Politics & Society 26* (3): 337–361.

Lindblom, Charles E. 1977. *Politics and Markets*. New York: Basic Books.

Lipschutz, Ronnie D. 1992. "Reconstructing World Politics: The Emergence of Global Civil Society." *Journal of International Studies 21* (3): 389–420.

Litfin, Karen T. 1994. *Ozone Discourses*. New York: Columbia University Press.

McAteer, Emily, and Simone Pulver. 2009. "The Corporate Boomerang: Shareholder Transnational Advocacy Networks Targeting Oil Companies in the Ecuadorian Amazon." *Global Environmental Politics 9* (1): 1–30.

McCright, Aaron M., and Riley E. Dunlap. 2003. "Defeating Kyoto: The Conservative Movement's Impact on US Climate Change Policy." *Social Problems 50* (3): 348–373.

McGregor, Ian M. 2011. "Disenfranchisement of Countries and Civil Society at COP 15 in Copenhagen." *Global Environmental Politics 11* (1): 1–7.

Meyer, John W., David John Frank, Ann Hironaka, Evan Schofer, and Nancy Branden Tuma. 1997. "The Structuring of a World Environmental Regime, 1870–1990." *International Organization 51* (4): 623–651.

Milner, Helen. 1991. "The Assumption of Anarchy in International Relations Theory: A Critique." *Review of International Studies 17* (1): 67–85.

Newell, Peter. 2000. *Climate for Change: Non-state Actors and the Global Politics of the Greenhouse*. Cambridge: Cambridge University Press.

Oberthuer, Sebastian, and Hermann E. Ott. 1999. *The Kyoto Protocol: International Climate Policy in the 21st Century*. Berlin: Springer.

Offe, Claus, and Helmut Wiesenthal. 1980. "Two Logics of Collective Action: Theoretical Notes on Social Class and Organizational Form." *Political Power and Social Theory 1*: 67–115.

Olson, Mancur, Jr. 1971. *The Logic of Collective Action: Public Goods and the Theory of Groups*. Cambridge, MA: Harvard University Press.

Pralle, Sarah B. 2003. "Venue Shopping, Political Strategy, and Policy Change: The Internationalization of Canadian Forest Advocacy." *Journal of Public Policy 23* (3): 233–260.

Pulver, Simone. 2002. "Organizing Business: Industry NGOs in the Climate Debates." *Greener Management International 39*: 55–67.

Pulver, Simone. 2007. "Making Sense of Corporate Environmentalism: An Environmental Contestation Approach to Analyzing the Causes and Consequences of the Climate Change Policy Split in the Oil Industry." *Organization & Environment 20* (1): 44–83.

Raustiala, Kal. 1997a. "The 'Participatory Revolution' in International Environmental Law." *Harvard Environmental Law Review 21* (2): 537–586.

Raustiala, Kal. 1997b. "States, NGOs, and International Environmental Institutions." *International Studies Quarterly* 41 (4): 719–740.

Reimann, Kim D. 2006. "A View from the Top: International Politics, Norms and the Worldwide Growth of NGOs." *International Studies Quarterly* 50 (1): 45–67. doi:10.1111/j.1468-2478.2006.00392.x.

Risse, Thomas. 2000. "'Let's Argue!': Communicative Action in World Politics." *International Organization* 54 (1): 1–39.

Ronit, Karsten. 2007. *Global Public Policy. Business and the Countervailing Powers of Civil Society.* Abingdon, UK: Routledge.

Rowlands, Ian H. 2001. "Classical Theories of International Relations." In *International Relations and Global Climate Change*, edited by Urs Luterbacher and Detlef F. Sprinz, 43–65. Cambridge, MA: MIT Press.

Schmidheiny, Stephan. 1992. *Changing Course: A Global Business Perspective on Development and the Environment.* Cambridge, MA: MIT Press.

Sell, Susan K., and Aseem Prakash. 2004. "Using Ideas Strategically: The Contest between Business and NGO Networks in Intellectual Property Rights." *International Studies Quarterly* 48 (1): 143–175.

Shawki, Noha. 2010. "Political Opportunity Structures and the Outcomes of Transnational Campaigns: A Comparison of Two Transnational Advocacy Networks." *Peace & Change* 35 (3): 381–411. doi:10.1111/j.1468-0130.2010.00640.x.

Sklair, Leslie. 1994. "Global Sociology and Global Environmental Change." In *Social Theory and the Global Environment*, edited by Michael Redclift and Ted Benton, 205–227. London: Routledge.

Smith, Jackie. 2004a. "Exploring Connections between Globalization and Political Mobilization." *Journal of World-Systems Research* 10 (1): 255–285.

Smith, Jackie. 2004b. "The World Social Forum and the Challenges of Global Democracy." *Global Networks* 4 (4): 413–421.

Spar, Debora. L. 1998. "The Spotlight and the Bottom Line: How Multinationals Export Human Rights." *Foreign Affairs* 77 (2): 7–12. doi:10.2307/20048784.

Strange, Susan. 1992. "States, Firms and Diplomacy." *International Affairs* 68 (1): 1–15.

Subsidiary Body for Scientific and Technological Advice (SBSTA). 1996a. *Mechanisms for Non-governmental Organization Consultations.* FCCC/SBSTA/1996/11. Bonn, GE: UNFCCC.

Subsidiary Body for Scientific and Technological Advice (SBSTA). 1996b. *Workshop on Consultative Mechanisms for Non-Governmental Organization Inputs to the United Nations Framework Convention on Climate Change.* FCCC/SBSTA/1996/MISC.2. Bonn: UNFCCC.

Tarrow, Sidney. 2001. "Transnational Politics: Contention and Institutions in International Politics." *Annual Review of Political Science* 4 (1): 1–20.

Tarrow, Sidney. 2005. *The New Transnational Activism.* Cambridge: Cambridge University Press.

United Nations Climate Change Secretariat. 2003. Non-governmental Organizations and the Climate Change Process. http://unfccc.int.

United Nations Climate Change Secretariat. 2014. Admitted Non-governmental Organizations. http://unfccc.int/parties_and_observers/observer_organizations/items/9519.php.

United Nations Framework Convention on Climate Change (UNFCCC). 1994a. A/RES/48/189. http://unfccc.int.

United Nations Framework Convention on Climate Change (UNFCCC). 1994b. A Mechanism for Dialogue Between FCCC Parties and Significant International Business Interests. A/AC.237/Misc. 43. Bonn: UNFCCC.

United Nations Framework Convention on Climate Change (UNFCCC). 1997. Mechanisms for Consultation with Non-governmental Organizations. FCCC/SBI/1997/14/Add.1. Bonn: UNFCCC.

Union of International Associations, ed. 2011. *Yearbook of International Associations 2009–2010.* Leiden and Boston: Brill.

Victor, David G., and Joshua C. House. 2006. "BP's Emissions Trading System." *Energy Policy 34* (15): 2100–2112.

Vogel, David J. 1996. "The Study of Business and Politics." *California Management Review 38* (3): 146–165.

Vormedal, Irja. 2008. "The Influence of Business and Industry NGOs in the Negotiation of the Kyoto Mechanisms: The Case of Carbon Capture and Storage in the CDM." *Global Environmental Politics 8* (4): 36–65.

Wapner, Paul. 1995. "Politics Beyond the State: Environmental Activism and World Civic Politics." *World Politics 47* (3): 311–340.

Watts, Michael J. 2005. "Righteous Oil? Human Rights, the Oil Complex, and Corporate Social Responsibility." *Annual Review of Environmental Resources 30*: 373–407.

Willetts, Peter. 1996. "From Stockholm to Rio and Beyond: The Impact of the Environmental Movement on the United Nations Consultative Arrangements for NGOs." *Review of International Studies 22* (1): 57–80.

Witter, Rebecca, Kimberly R. Marion Suiseeya, Rebecca L. Gruby, Sarah Hitchner, Edward M. Maclin, Maggie Bourque, and J. Peter Brosius. 2015. "Moments of Influence in Global Environmental Governance." *Environmental Politics 24* (6): 894–912.

Yamin, Farhama, and Joanna Depledge. 2004. *The International Climate Regime: A Guide to Rules, Institutions, and Procedures.* Cambridge: Cambridge University Press.

Yergin, Daniel. 1992. *The Prize: The Epic Quest for Oil, Money, and Power.* New York: Simon & Schuster.

# Part III

# Expertise and informational economies

# 7 Preventing the resource curse

## Ethnographic notes on an economic experiment[1]

*Gisa Weszkalnys*

Sometime in 2002, an email was sent from São Tomé and Príncipe (STP), a small island state off the West African coast, to the renowned US economist Jeffrey Sachs who counts "shock therapy" in the former Soviet Union and the UN Millennium Development Goals among his best-known feats. The email asked for Sachs's help in saving the tiny nation from becoming the victim of international oil business. Its sender was a former employee of the US State Department, Doreen,[2] whom I met when conducting fieldwork there in 2007.[3] Doreen was now the local representative of Sachs's project in STP. She told me how, when disputes regarding STP's speculated oil resources first began in the early 2000s, the newly installed president Fradique de Menezes had called on her with the intention of preventing the country from social and political disintegration. Having retired to STP some years earlier, Doreen was closely acquainted with many of the Santomean Who's Who. She had a reputation as an informal advisor, occasional translator, and was known as somebody who could solicit foreign advice because of her connections to, and understanding of, the world of African and international politics and diplomacy. Doreen recalled that Santomeans did not want to repeat the experience of their neighbors. She mused that the isolated island state situated in the "bad" neighborhood of the Gulf of Guinea was ill equipped for the sudden encounter with the cutthroat world of oil, just like native peoples elsewhere, and had no one there to defend them. The story seemed to have intrigued Jeff Sachs who, always on his Blackberry, replied straightaway: "Tell me more."

This chapter examines how high-powered economic theory, like that produced by Jeffrey Sachs, plays out in an emergent African oil economy. I am particularly interested in the ways that Sachs's work in STP (which he began subsequent to the email exchange above) articulated the idea of the "resource curse," a specific economic device of which Sachs has been a major proponent (Sachs and Warner 1995, 2001; see also Auty 1993; Karl 1997). My analysis of Sachs's work resembles Timothy Mitchell's memorable account of Hernando de Soto and the Urban Rights Project in Peru and has similar intentions. The analysis of these kinds of economic experiments "illuminate[s] the relationship between economics and the object it studies" (Mitchell 2005, 297). In other words, it demonstrates how economics enacts its worlds. However, my chapter is not just an illustration of the process of economization, that is, "the contribution of economics to the constitution of the economy," which Çalişkan and Callon have recently discussed (2009, 370; 2010). Rather, it adds a critical perspective on what are often quite sanitized accounts of what happens when economic theories are implemented or enacted in specific contexts. Such accounts focus on technical detail but leave out the messy realities of historical, cultural, social, and political conditions with which the theory inevitably articulates.

I argue that economic tools and theories are not simply imposed or implemented in different contexts but rather collide with pre-existing conditions and result in complex articulations. Current approaches to economics developed by sociologists and anthropologists over the last decades have opened up a mode of analyzing economics in ways not prefigured by conventional economics but also not explicitly anti-economic (Barry and Slater 2002, 180; see also Callon 1998; Carrier and Miller 1998; MacKenzie, Muniesa, and Siu 2007). This work has explored the role played by economic theory and economics, more generally, as a discipline and practice, a set of ideas and technologies, in the constitution of economies. It shows how economics shapes the behavior of the economic agents that are its subjects.[4] The contention is that economics creates or "performs" the worlds that it professes merely to describe. Mitchell's study of de Soto's Peruvian experiment is a great illustration, in this instance, of an attempt to reformat an entire country's economy. It also offers a critical alternative interpretation of its results (Mitchell 2005). Operating through his Institute for Liberty and Democracy, de Soto is the central character in this undertaking—a supposed "neoliberal from the third world" (Mitchell 2005, 306) who had been discovered by Friedrich Hayek in the late 1970s. His Urban Property Rights Project sponsored by the World Bank claimed to turn around people's lives by giving them more economic opportunities through the regularization of private property rights in urban neighborhoods. While the property titling did not, as intended, lead to an increase in credit to poor Peruvians, it had an unintended side effect as the new property owners apparently began to work more. What Mitchell shows is how the world of de Soto's "natural" experiment had already, to some extent, been carefully staked out and rearranged to then yield the desired observations.

Mitchell's account is an important critique of neoliberal interventions into so-called developing economies (see also Klein 2007). It demonstrates how these interventions are rarely straightforward but are what Callon (1998) terms complex per-formations that draw on projects, programs, policies, and other socio-technical arrangements, which have already shaped the world in such a way as to make the experiment possible. However, what Mitchell's account does not attend to in as much detail are the reverberations of such experiments in local discourses, the ways in which these experiments get folded into existing and reformed practices and institutions, how they are sometimes contested, and how they often draw on pre-existing cultural perceptions of economic worlds. For example, we hear little about what Peruvians made of de Soto's experiments (and of other neoliberal programs that were reshaping their world), or whether they began to interpret their own actions in the terms offered by global technocrats.

Economic experiments or devices are not simply imposed on existing or pre-arranged worlds; in becoming preferred framings of these worlds, they also have to adjust to and articulate with them. Here, I use STP as a case that provides a compelling illustration of the multiple articulations of the resource curse thesis. Following the announcement of potential oil resources in the late 1990s, STP set out to turn itself into an exemplar among African oil-rich states. The intensity of the interest paid to STP, particularly by the international financial institutions (IFIs), NGOs, policy advisors, and international experts, has been quite extraordinary, considering that to date oil has had but a speculative presence.[5] While conducting fieldwork in STP, I was stunned by the concrete and material effects that the expectation of oil has had at a time when not a drop of it had yet been extracted. These include a new building for a National Petroleum Agency, an extensive World Bank governance and capacity building program, latrines and wells sponsored by Chevron Texaco, and numerous campaigns and conferences by international NGOs seeking to implement new modes of democratic and corporate accountability. This broad-based interest is partly related to STP's small size—1001 $km^2$ with barely 200,000 inhabitants[6]— and, no doubt, to its location in the Gulf of Guinea, an oil-rich region which has been at the heart of much recent scholarly and policy debate about the effects of oil (e.g., Soares de Oliveira 2007; Obeng-Odoom 2015; Weszkalnys 2009).[7]

This debate pivots largely on the notion of the "resource curse," a notion which implies, briefly, that natural resources—and oil seems to stand out— have failed to deliver prosperity to their African producers and, instead, impact negatively on their economic development. This seems to go hand in hand with an increase in corruption, conflict and violence (see Rosser 2006b; Ross 2015). While these accounts avoid a simplistic resource determinism (cf. Watts 2004), they construe oil in particular as a resource with unique qualities and characteristics: it is finite and highly valuable but, generally, a volatile source of income; its extraction is capital intensive and requires specialized skills; and it is said to have socially and politically disintegrative qualities, sometimes even when it is only a future expectation. And whereas earlier literature was puzzled by the resource curse effect, more recently, there have been various attempts to explain, and eventually cure, the curse (e.g., Humphreys, Sachs, and Stieglitz 2007; Rosser 2006a). To mitigate oil's effects, some economists now suggest, quite literally, that oil economies be dispersed, for example, by distributing oil revenues among citizens in order to "simulate a situation in which the government has no easy access to natural resource revenue, just as governments in countries without natural resources" (Sala-i-Martin and Subramanian 2003, cit. in Pegg 2005, 24; see also Ferguson 2015).

Scholarly and journalistic accounts of STP's prospective oil boom are steeped in ambiguity. They ask whether oil might turn the small island state into the next Kuwait and they worry that the symptoms of the resource curse are already manifesting themselves (e.g., Shaxson 2007; Vicente 2010). They even envision a version of the Niger Delta in STP's tropical forest (Peel 2009, prologue). The resource curse has also been actively mobilized in the context of technical programs, research projects, and NGO campaigns. In my conversations with local administrators and professionals, with World Bank staff, UNDP employees, and NGO workers, I was almost invariably offered assessments of STP's current situation and potential future in terms echoing the academic literature. Although all hoped oil would make a positive difference to STP's economic stagnation, they expected it to bring first and foremost political instability, social friction, and serious misconduct on the part of political elites.

This chapter aims to demonstrate that, in STP, the resource curse has come in various guises: as an economic concept, a narrative tool, a policy instrument, an abstraction, a future imaginary, and so on. My argument is two-fold. First, I am interested in the agentive qualities of the resource curse and the practices it enables, that is, the likely and unlikely futures it has called forth, and how it has spurred people into action. Second, rather than understanding the activities of advisors, technicians or civil society activists as mere performances of the curse, I argue that they are also the material sites of its production and contestation. This multi-faceted economic device articulates with other instantiations of itself, with local history and cultural understandings, and with political exigencies, through which it has been actualized and transformed.

## An economy of expectations

What circulates in STP today, instead of oil, are the institutions, expert documents, IMF reports, and discourses on transparency and corruption, in other words, the infrastructures of "ethical capitalism" (Barry 2004) that are designed to ease the increasingly controversial production of oil and the distribution of the wealth it generates. Although these infrastructures aim to make resource extraction more just, they also help to construct a new *Afrique utile*, that is, to reduce risk and provide relative stability and security, which will allow capital investment and extraction to occur uninterrupted (cf. Reno 1999, in Ferguson 2006, 39). Jeff Sachs's economic experiment in STP was constitutive of this context.

During the first decade of the new millennium, the World Bank invested millions of dollars into a governance and capacity building program in STP, including the implementation of infrastructures, institutions, and technical and ethical know how aimed at facilitating the management of future oil resources. The London-based NGO International Alert (IA), supported by UNICEF and USAID, set up a media center for local journalists and two community radio stations. Together with the Publish What You Pay Campaign (funded by philanthropist George Soros) and the NGO Global Witness, IA held two conferences in STP to which civil society groups from the Gulf of Guinea region were invited. In addition, a mixed group of parliamentarians, local business people, journalists, and other select representatives of Santomean civil society was taken to Norway to experience a country often held up as an exemplary oil economy that has managed to escape the curse. In a further bid to constitute itself as a transparent oil state, STP has endorsed the UK-led Extractive Industries Transparency Initiative (EITI) and, with subsequent funding, has installed a national committee that convened government, industry and civil society representatives.[8]

Over the few last decades, STP has undergone a transformation from a derelict plantation economy into what I term an economy of expectations— a transformation which has pivoted largely, though not exclusively, on the prospect of oil. From the sixteenth century, the Portuguese established a flourishing colonial economy on the islands based on slavery and imported labor from Angola, Cape Verde, and other parts of Africa, which produced sugar, coffee and cocoa. This economy faltered in the early twentieth century, as it came under attack from British reformers lamenting the continued use of forced labor, on the one hand (Satre 2005), and the vagaries of a shifting global cocoa market, on the other. After independence in 1975, the already unprofitable plantations were nationalized while mismanagement and falling cocoa prices soon led to their virtual collapse (Frynas, Wood, and Soares de Oliveira 2003; Seibert 2006, 45). In the late 1980s, STP became the second African state to adopt multiparty democracy (Seibert 2006, 1). A structural adjustment package revolving around the privatization of land brought uneven but overall disappointing results. Few of the former plantation workers possessed the skills, the will, or the capital to work the land as independent farmers. In addition, some of the best land was quickly allocated on a preferential basis to influential members of Santomean society. The majority of Santomeans, today, live off subsistence farming, fishing, and petty trade. Cocoa still represents the vast majority of all exports, and there is no other export industry to speak of.

In this context, the idea that oil could function as a savior had an astounding grip on popular and scholarly imaginations of STP. STP's speculative oil future emerged from a meeting of minds and changing fortunes in the outback of the capitalist world. In the late 1980s, an Angolan-based explorer and entrepreneur, Christian Hellinger, started drilling for oil on the islands.[9] At the time, the Gulf of Guinea region had been enthralled with a new oil boom as a consequence of significant finds in Equatorial Guinea and technological advances in ultra-deep sea exploration, which promised considerable returns. Hellinger had been using STP as a retreat for his staff from his diamond mines in war-torn Angola. He built a hotel, started a construction company, and cultivated excellent relations with the Santomean government. However, despite a multi-million dollar investment, he never struck oil and instead sold the geophysical data obtained to the Environmental Remediation Holding Company (ERHC), a small US company with uncertain credentials. STP went on to enter agreements with ERHC and the multinational Mobil (later

ExxonMobil). Both companies were guaranteed rights to exploration and revenue shares, which are now deemed to have vastly exceeded what is standard in the industry (Shaxson 2007, 156).

In the story about its oil, the small island state is often cast as a victim— ignorant at worst, naïve at best. Nigeria, STP's huge, savvy neighbor, took advantage of the country by initiating a maritime border dispute. This was eventually settled through the establishment of a Joint Development Zone (JDZ) of which Nigeria holds 60 percent and STP 40 percent. It is governed by a so-called Joint Development Authority that has its main seat in Abuja. Similarly, the two licensing rounds for the JDZ, run in 2003 and 2005, were heavily disputed. The results of one of these rounds were called into doubt because of the participation of too many small companies with Nigerian connections, little technical experience, and dubious objectives.[10] More than a decade later, STP is still not producing any oil. In March 2010, STP opened the long-awaited first licensing round for its Exclusive Economic Zone (EEZ), reinforcing the impression that there is (still) an oil economy in the making (Weszkalnys 2015). As of the writing of this paper, however, oil exploration in STP has not shown any significant positive results. STP's former president, Fradique de Menezes, who held office for two terms between 2001 and 2011, is generally viewed as having been instrumental in pursuing the issue of accountability in the oil sector. His reputed acceptance of US$100,000 from ERHC for an election campaign (Seibert 2008, 125; Soares de Oliveira 2007, 238) has been but one of many apparent paradoxes in this speculative oil state, which has made such great efforts to cast itself as prudent and transparent.

The mention of vast natural resources alone may cause the imagination to run wild. Anna Tsing (2000) beautifully illustrates this in her account of an Indonesian gold rush where speculation and conjecture may be sufficient to produce an "economy of appearances," which grows and grows until its spectacular collapse. STP's oil economy without oil has depended on similar modes of dramatic and economic performance to attract potential investors, as "profit must be imagined before it can be extracted" (Tsing 2000, 118). Oil could be speculated into existence, substantiated by material claims. Mystery and rumor were equally part of the story, only that, here, the dominant rumor concerned suspected cases of corruption, a rumor not unfamiliar in the small island state, but now cast in a new guise and with potentially high sums at stake.

In other words, what made this situation distinct was the simultaneous elevation of both positive (of money, wealth, and development) and negative expectations right from the start. There is a flipside to the expectations of the risk-taking businessmen and speculators of Hellinger's caliber who similarly populate Tsing's account. STP's universe of expectations also includes those of the local population, political elites and international advisors. There is the hope of betterment and wealth of people who get by for the most part on a few dollars a day. There is the anticipation of those in power that oil money will kickstart long-awaited development. And there is the expectation of people who see opportunities for personal aggrandizement arising in the shadow of the new oil economy. There are also the failed expectations of those who thought their education would give them a head start in the budding oil economy, and who find that there is just not enough oil-related work to go around. Finally, there are the expectations of an incipient resource curse entertained by the economists, NGO representatives and others who are the focus of this chapter.

Framed by the curse, people's hopes and expectations have become an object whose circulation is carefully observed and guarded. Hopes and expectations associated with oil are perceived as troublesome: a valued but potentially dangerous good.[11] They are often assumed to be exaggerated and illusionary, and to fuel the curse. To STP's budding oil economy, hopes and expectations are negative externalities (Callon 1998). They are fuzzy, incalculable and hard to control, they may spring up at any moment, and often appear thoroughly irrational. Sociologically speaking, they are interests and factors that emerge from but exceed and "overflow" the frame set up by market transactions, and (in contrast to positive externalities) cannot be tamed and put to use. Quite the contrary: hope, as an externality to STP's emergent oil economy, is perceived as a threat to its future development. On numerous occasions IFI staff and some of my local interlocutors who closely follow the Santomean situation had the following to say to me about this issue. On the one hand, complacency about imminent wealth may induce passivity and laziness in the population—a kind of "cargo cult" attitude—that would be detrimental to an already badly performing economy. On the other hand, hopes for effortless wealth and development are deemed problematic because they may lead to increasing demands and pressure on government for social provisions, higher salaries, and the like, demands that it cannot (or should not) fulfill as this might lead to imprudent overspending. In either case, economic instability and social conflict may arise.

In this economy of expectations, the activities mentioned at the start of this section, conducted by international experts, transnational agencies, citizens, and the national government, aim to eliminate these externalities or at least to make them manageable. They aim to turn Santomeans into rational citizens, ethical leaders, civic-minded politicians, forward-looking state servants, and resourceful entrepreneurs. They try to put into place appropriate conditions (technical, legal, institutional, behavioral, etc.) in anticipation of a potential resource curse.[12]

## STP as an economic experiment

"A natural experiment in economics," as Mitchell suggests, "is not an experiment carried out in nature. It is an establishing of facts carried out in a world that has been organized to make it possible for economic knowledge to be made" (2005, 304). One particularly remarkable experiment of this kind was the advisory project led by Jeff Sachs, which emerged from the exchange of emails described at the beginning of this chapter. Following this exchange, a meeting was arranged between Sachs and STP's president de Menezes in New York where they discussed the country's future. The rest is history. With sponsorship from the Open Society Institute (another philanthropic venture set up by multi-billionaire George Soros),[13] Sachs pulled off an impressive project involving a team of high-powered lawyers and influential economists, all working *pro bono*, and a bunch of bright graduate students from the Earth Institute at Columbia University (which he directs). On repeated trips to the islands beginning in 2003, Sachs's team devised a holistic approach to STP as a future oil state, taking into account aspects as wide-ranging as malaria, sanitation, electrification, and the law.[14] They sought to design a whole socio-natural-technical arrangement that would provide the ideal conditions for oil exploitation in STP and save this (otherwise) "doomed" nascent oil economy.

Neo-classical economics has been characterized by a methodological individualism, an inability to conceive of society as something more than an aggregate of individual behavior and an inability to explain what "the social" is all about (Fine 1998, 50). More recently, the discipline can be seen to be moving in two opposed directions. On the one hand, there is the attempt to apply the principles of neo-classical economics to everything, including all non-market behavior (Fine 1998). On the other, I suggest we are witnessing a renewed interest in a kind of "socially responsible" economic knowledge that seeks an explicit and direct engagement with those messy socio-political worlds in which economics are embedded.[15] Sachs's STP project seems to fit this picture.

Sachs's current work displays a deep concern with poverty alleviation (Sachs 2005a); the solutions he puts forward hinge on geographic and environmental factors as well as economic institutions, notably capitalism (see also Sachs 2000),[16] and ascribe a unique role to universities (Sachs 2005b). This may seem like an about-turn from the projects that made him a globally renowned economist. Sachs has been branded one of the vanguards of neoliberal reform and "shock therapy" which are presented as having led to an improvement of economic development and general living conditions in the developing and former socialist world. Instead, as Naomi Klein reports, there are numerous instances where shock measures resulted in harsher conditions and occasionally outright violence against citizens (Klein 2007). Responding to these claims in an interview with Klein, Sachs implied that his advice to the Bolivian president Victor Paz Estenssoro in the 1980s and to Boris Yeltsin in the 1990s was full of good intentions, but that inexperience and a degree of political naivety prevented him from anticipating the real consequences.

From this perspective, Sachs's involvement in STP might simply appear to be a revision of those earlier projects. His co-editorship with Joseph Stiglitz (an outspoken critic of World Bank and IMF free-market policies) of a recent collection on how to "escape the resource curse" (Humphreys, Sachs, and Stieglitz 2007) could be seen as further evidence of his attempt to pursue a more tamed, philanthropic neoliberalism. Francesco Guala's account of the new experimental economics, another reaction to the apparent shortcomings of conventional economic theory, is helpful here (Guala 2007). Guala distinguishes between "testers" and "builders." Builders, in a simplified sense, are those economists who—unlike testers—do not stop at testing and simply noting the failure of economic theories. Instead, they continue to ask what institutions or rules would need to be introduced to make their subjects fall "in line" and to make experimental results correspond to the theories. Sachs can be seen a builder in this sense.

Sachs's advisory team conceived STP as a kind of real-world experiment whose central objective was to develop a framework for transparency in public expenditure as well as Plan of Action for sustainable economic development. Thanks to them and their collaborating Santomean lawyers, STP now has comprehensive, state-of-the-art legislation governing its oil affairs (Bell and Faria 2007) that was drafted in collaboration with two Santomean lawyers.[17] With this law, enacted in December 2004, STP received a National Oil Account and a Permanent Fund for future generations, as well as an Oversight Commission and a Public Information Office, which were finally installed in 2009.

The team's efforts were much appreciated and delivered a prestigious project that helped the country demonstrate its willingness to enact measures of good governance. However, it is unclear to what extent the legislation and related measures will be effective. Although the models for transparency and accountability may be made to travel, they are also readily inflected by their new surroundings (cf. Reyna 2007). The Earth Institute project ground to a halt in 2007 except for some continued advisory work in the domain of environmental legislation and the compilation of an investor's guide to STP (Earth Institute 2008). President de Menezes's relations with his academic friends in New York became increasingly frosty. For example, when de Menezes visited New York in 2008 to attend the UN Assembly, Sachs arranged for a formal dinner with around 20 guests. When the president finally arrived, with a customary 2-hour delay, Sachs had apparently already left (Seibert 2010). More generally, international observers have worried about the apparent reluctance of Santomean political authorities to subscribe to transparency mechanisms, reinforced by the country's expulsion from the EITI process in 2010 (Vines 2010a). However, local explanations for the failure to meet EITI criteria pointed to a lack of technical and logistical capacity. Two years later, the country was readmitted with substantial support from the World Bank, a new committee was installed, and two reports detailing STP's oil-related revenues have been published.

## Opening the black box

But what is this resource curse of which STP has come to be seen as an instantiation even without oil? Why has it provoked so much activity in the small island state? From the late 1980s, resource economists and political scientists began commenting on the surprising observation that the presence of natural resources does not automatically translate into overall economic development. Richard Auty, for example, noted that "a favourable resource endowment may be less beneficial to countries at low- and mid-income levels of development than the conventional wisdom might suppose"; indeed, they "may actually perform worse than less well-endowed countries" (Auty 1993, 1; see also Gelb and Associates 1988; Sachs and Warner 1995, 2001). In addition to economic under-performance, the presence of mineral resources—particularly oil—came to be associated with social, political and ethnic conflict, and civil war (Collier and Hoeffler 2000; see also Behrends 2008). By developing a new economic device scholars could begin to explain surprising and, to some extent, counter-intuitive observations (Rosser 2006a, 2006b). Conventional development economics assumed the presence of resources to be a beneficial trigger of development; and a Malthusian paradigm would associate resource *scarcity* rather than abundance with social conflicts and wars (see also Peluso and Watts 2001).[18] The resource curse challenged this assumption.

Put differently, the concept of a resource curse responded to an apparent failure of economic theory. For economists, the resource curse initially posed a puzzle. Their explanations—including "a decline in the terms of trade for primary commodities, the instability of international commodity markets, the poor economic linkages between resource and nonresource sectors, and an ailment commonly known as the 'Dutch Disease' " (Ross 1999, 298)—seemed to leave clear gaps. Meanwhile, political scientists, policy researchers, and nongovernmental organizations began to add cognitive, societal, and state-centered explanations, emphasizing the dominance of state-owned enterprises and the weakness of state institutions (e.g., Karl 1997; Ross 1999). In 1999, London-based NGO Global Witness made path-breaking revelations. Its notorious report *A Crude Awakening* uncovered the ways in which Angola's oil financed the country's protracted civil war, and highlighted how the sinister practices of governments, rebel groups, and the activities of international oil companies were deeply entangled (Shaxson 2007).

During the same period, the Berlin-based NGO Transparency International had begun to publish its now much-cited transparency indices. The IFIs had moved the fight against "corruption" to the top of the their agendas. Whereas corruption was once seen as a result of poverty, beginning in the 1990s, the Washington Consensus became that corrupt countries are poor (Krastev 2004). As corruption and mismanagement came to be seen as key causes of economic (under)performance and poverty, bad governance rather than resources *per se* emerged as the trigger of the curse (Leite and Weidmann 1999). The resource curse has now become the dominant narrative, particularly in respect to developing oil states. Oil is considered the most problematic resource for various reasons, including its specific qualities as a non-renewable resource whose extraction requires huge capital investment and specialized skill. Oil is also a highly lucrative resource with the apparent capacity to increase rent seeking and to decrease state actors' need to answer to their citizens. Meanwhile African oil producers in particular find themselves branded as extreme cases of political, economic, and ethical impropriety and as surprisingly "successful failed states" (Soares de Oliveira 2007).

As calls to act on the resource curse theory become progressively louder, its enactments proliferate. Let me mention just two important examples. In 2002, Tony Blair announced the EITI at the World Summit for Sustainable Development in Johannesburg. This was an early effort to create transparency by bringing national governments, companies and civil society to one table with open books. Since then, the World Bank has initiated new projects in which it claimed to use resource exploitation for poverty alleviation and development. The Chad-Cameroon pipeline is but one prominent (and failed) project of this kind. The so-called "Doba model" (after the Doba oil field in southern Chad) was to be "a pioneering model for responsible private investment in Africa and the developing world" (Massey and May 2005, 255), complete with an Oil Revenues Management Act, an Oil Account, an Oversight Commission, and an International Advisory Group (see also Pegg 2005; Reyna 2007). In reality, the Doba model quickly appeared to be nothing more than "a template for the oil industry and host governments that marks out the minimal requirements necessary to secure the World Bank's approval for future projects" (Massey and May 2005, 274). Despite the measures taken, corruption persisted and money was spent for purposes outside the revenue management plan (including weapons). In addition to negative environmental and social impacts, far from having contravened the resource curse in Chad, the pipeline project has brought disillusionment among the local population regarding the promises of "good governance" and collaboration between the diverse parties involved (see also Leonard 2016).

In other words, the resource curse works like a black box in resource economics. It explains what common sense and established theory seems unable to explain. As a concept and mode of explanation, it "works" but nobody quite knows why or how. In science and technology studies, the black box has been a handy analytic device that refers to a technological complex whose functioning is deliberately left obscure; only input and output matter (Latour 1987; Winner 1993). Interestingly, the black box of the resource curse conceals not a functioning but a failure. This is the failure to explain certain observations with the data and tools commonly available to economists.

During the last decade the concept has become subject to greater scrutiny. Some scholars have interrogated the association between resource dependence and slow economic growth (Brunnschweiler and Bulte 2008), the impact of natural resources on democracy (Haber and Menaldo 2010), or the relationship between oil and conflict (Rigterink 2010). These studies offer a sophisticated qualitative and statistical un-making of the resource curse thesis, and an argument that what may be at work, here, is correlation rather than causation (see also Rosser, 2006b). Without doubting the existence of a curse, Ross (2015) has invited researchers to examine the precise conditions generating a curse in one locale rather than another. Experts of the "social"—including anthropologists and sociologists—are occasionally called to task to reveal the curse's inner workings and components. Others have similarly pointed to a need for more nuanced historical and sociological studies taking account of existing structures of power into which oil is inserted (Reyna and Behrends 2008; Watts 2004). They show how oil can exacerbate already existing political conflicts and substitute for preceding opportunities for personal enrichment (arms, diamonds, development aid, etc.) by political and economic elites (e.g., Massey and May 2010; Seibert 2008; Vines 2010b; Watts 2001, 2004). Accounts of rent-seeking behavior, corrupt elites, political culture, and colonial legacies have begun to fill the box.

This could also be observed in the STP advisory project. By providing the foundations for possible future social and economic policy in STP, Sachs and his team had to open the curse's black box and craft some content for it. This involved drawing on foreign (non-economic) experts' understanding of comparable cases and local experts' experience of the Santomean reality to buttress the economists' insight. Moreover, the input of a wide range of "stakeholders" was hoped to add up to a cure for the curse. This approach—not unusual in the context of international development (and espoused, for example, in the Millennium Project also coordinated by Sachs)—claims to respond to local conditions while offering models that can be replicated globally.

In some sense, the resource curse is an economic device par excellence of the type examined by Michel Callon and other sociologists of the economy in recent years. Their studies have shown how such devices are enacted and performed, thus creating their own worlds. However, there is a limitation to this type of analysis in that it occasionally overlooks the complex ways in which the device—once enacted—itself undergoes a transformation. For example, when looked at ethnographically, the resource curse is not simply an economic instrument that emerged in the research centers of World Bank and IMF or in the academic ivory towers of Harvard or Columbia, and was then exported to, and enacted in, the context of developing countries. Although I do not want to diminish the intellectual work that went into developing the resource curse thesis, I suggest that the story of its invention I told above is but one possible scenario. We could imagine instead, that the curse might initially have been a kind of para-economic observation (to paraphrase Holmes and Marcus 2005). It may have been an idea that took the form of tales and anecdotes circulating among civic activists, foreign correspondents, and industry consultants stationed in Venezuela, Nigeria, or Angola before entering the academy, NGO campaigns, and World Bank jargon. Dressed up with economists' statistics and reframed econometrically, these ideas became officially palatable to the technocrats of the IFIs.

In other words, devices such as the resource curse unfold in multiple ways, articulating with already prevalent ideas about past, present, and future of economic development. The resource curse provides a powerful purchase on the direction and shape that economic—and associated political and social—development will take. Yet it gets entangled with other existing imaginaries. It is a narrative device that entices, and speaks of money, greed, power, and mysterious unknown forces. It is persuasive as a simple explanation of present-day events, and is readily adopted and adapted to what people already know. It is these kinds of complex articulations to which I now turn.[19]

## Articulations of an economic device

I want to discuss four, quite different instances of what I term the articulation of the economic device. The first shows how economic theory comes to frame empirical observation, leading to interpretations of individual events in pre-determined but contentious ways. The second form of articulation is the economic experiment of Sachs's advisory project. My discussion focuses on its effort to engage the negative externality of hope with its program for a cure to the curse, and the concomitant reconfiguration of the Santomean "economic subject" (Roitman 2005). That is, how did the project seek to effect a change in people's place in relation to STP's reformed political economy with its shifting practices and significations of good and bad oil wealth and ethical comportment? The third instance is a Santomean friend's poignant commentary on the notion of "hope" in the African context, which led me to reflect on the kind of economic imaginaries that over time have been invoked in the small island state. How is the expectation of oil entangled with other expectations regarding economic growth in the islands? Lastly, I address the question of how the resource curse can become constitutive of individual biographies by shaping economic outlooks and life trajectories. Together, the four examples also constitute a tentative critique of the ability of the resource curse thesis to make sense of local apprehensions of the past, present and future consequences of extractive industry developments.

## *I*

In the busy activities carried out by international NGOs, IFIs, scholarly experts, and independent advisors in and about STP, the country has come to be an instantiation of a particular "African" phenomenon. It is an example of yet another country threatening to mis-develop once resources are exploited. When resource wealth appears on the horizon of possibility, so the story goes, political elites become greedy. In the words of one of the lawyers who worked on Jeff Sachs's Santomean advisory project quoted in a New York Times article: "In West Africa, the scent of oil alone may be enough [to produce corruption]."[20]

Even though commercial oil extraction continues to be delayed, for some observers the resource curse has already arrived in STP. In this view, oil has significant agency and is able to produce instantaneous social and political effects. For instance, studies describing the country's emergent oil economy have focused on a coup d'état in 2003, which is portrayed as out of place in what commentators and Santomeans alike are wont to romanticize as a peaceful idyll.[21] There is some agreement that the coup was caused by the greed and exaggerated expectations welling up in anticipation of oil (e.g., Sandbu 2004;

Vicente 2010). A local rebel leader is cited as claiming: "Oil was at the bottom of this. Yes. It's oil" (Shaxson 2007, 160). In short, oil caused the coup.

From a different perspective, however, oil seemed to be just another, albeit extremely powerful, factor in an already grievous and conflicted socio-political environment. As Gerhard Seibert (2003) has shown, the 2003 coup was not a first. In 1995, some years before "oil" became an issue in STP, rebels arrested the president Miguel Trovoada in protest against widespread corruption and the worsening conditions under which the military had to work. Order was reestablished after a week, although political instability continued to plague the small country, which has had a long series of successive governments in the last two decades. The 2003 coup followed a similar political crisis. In January, after the majority of his party's MPs called off their support for him, president de Menezes dissolved parliament and ordered early elections. Although nego-tiations quickly led to a restoration of the parliament, this occurred in a context of growing popular discontent with the country's fragile social and economic situation. In April 2003, groups of citizens undertook repeated civic actions to complain about corruption and a lack of transparency in government circles. This culminated in a demonstration during which police shot one of the protesters.

In July 2003, a group of soldiers took control of government and public buildings and arrested several ministers after alleging corruption. They said they were acting in protest against the socio-economic misery and demanded improved conditions for the armed forces. The group that instigated the coup was the Christian Democratic Front (FDC), an alliance of former Santomean resistance fighters who had attempted to overthrow the socialist government in the late 1980s and a group of Santomeans who had fought as part of the South African 32-battalion (the "Buffalo Battalion") in the Angolan civil war (some of whom later joined the infamous private mercenary force Executive Outcomes). On returning to STP in the 1990s, members of the FDC sought compensation and support for their "reintegration into society" from the government. Having been denied this in 2003, they launched a coup. The coup only lasted a few days and was resolved with the help of international mediation and the signing of a memorandum of agreement between the president and the rebels which made specific reference to the latter's demands.

As Seibert's erudite discussion demonstrates, oil "cannot be considered [the] principal cause" of the coup (Seibert 2003, 10), during which existing grievances of political opposition aligned with those of the military. Oil certainly played a role in mobilizing international organizations and foreign governments, such as Angola, Brazil, Gabon, Mozambique, Nigeria, Portugal, South Africa, and the US, to get involved in the post-coup negotiations. But, instead of viewing the coup as *caused* by oil, I would argue it is more apt to say that the coup *came to matter* because of the expectation of oil.

My interview with the president of the local Adventist Church, Rev. José Dias Marques, confirmed this point. In his view, the coup had been an expression of prevalent political factions and bitterness about social inequities on the islands. Oil had simply been injected into this explosive concoction. At the time, Dias headed the executive committee of a National Forum that was conceived in the memorandum of agreement between government and coup plotters as a solution to the social and political tensions expressed by the coup. However, instead of solving these, Dias recalled that the National Forum had become almost exclusively focused on oil. Under the influence of the advisory project headed by Jeff Sachs (which had started its work in STP at roughly the same time), the Forum became a kind of nation-wide awareness campaign about the consequences of oil. A Santomean UNDP employee with whom I discussed the Forum was similarly skeptical. Although the Forum was celebrated as exemplary, she claimed that in actual fact people didn't read the public information brochures that were handed out and paid little attention to the public discussions because they were so upset and disillusioned about the lack of development and about continuing social inequities and political instabilities over the last decade.

The assumption that the discovery of oil constitutes a watershed event, characteristic of a great deal of economic and political analysis of "the curse," makes history and culture appear largely negligible. Oil is said to provoke a sudden rupture, effecting negative changes in economic development and in people's behavior at an individual and a group level. It becomes all too easy to overlook, as Watts (2004) notes in relation to the Niger Delta conflict, that in such contexts oil is frequently merely a new "idiom for doing politics . . . inserted into an *already existing* political landscape of forces, identities, and forms of power" (76, my emphasis). Protracted local histories of greed and grievance, however, fall by the wayside in resource curse narratives. So too, do more long-standing regional historical experiences of resource exploitation such as those found under colonial rule where mercantilist and prebendalist tendencies occasionally show uncanny resemblance with contemporary "rentier" state practices (cf. Omeje 2008, 2–4).

## II

The notion that the announcement of oil constitutes a watershed event in STP's history also underpinned the Earth Institute advisory project. Sachs's team, as we have seen, was key in making oil an explicitly public issue. Taking advantage of the National Forum (the purpose of which was to bring unity to the country destabilized by military and social unrest), deliberative meetings were held across the islands' former plantations, in villages, and in the capital. A broad range of people answered questionnaires that assessed their expectations regarding a future with oil. The campaign envisioned a future crisis, whose prevention requires the adoption of certain strategies, including institutional and technical capacity building, the implementation of a watertight legal framework, and what might be termed the creation of certain ethical "regimes of living" (Collier and Lakoff 2005). The analogy invoked to persuade Santomeans, who had no previous experience of oil, was that of a flood. In one of the leaflets handed out to the population as part of the public information campaign run by Sachs's team, cartoon-type images showed a country submerged and completely destroyed by a big flood. As a positive contrast, there were images of a country safe and flourishing because a dam had protected it from devastation. The dam was the oil law that the advisory team was about to propose: "Imagine what would happen if there was a big flood that hit us . . . The oil law creates a dam."[22]

A parallel with the shock therapies in Chile or Bolivia does not seem too far-fetched (cf. Klein 2007). In STP, invoking an imaginary future "shock" similarly justified a need for action, creating a playground for economists, lawyers, and allied scholars to try out their ideas, models, and strategies in a "real-world" experiment. Arguably, the strategies that Sachs and his team proposed for STP are not as harsh as those of earlier shock programs. There were no austerity measures, no cuts in public spending, no privatization (some of this was already in place). Instead the team presented a rounded Plan of Action for sustainable development (Sachs et al. 2005), which included suggestions for the improvement of agriculture and tourism as much as public health (notably in regards to malaria). Investment in public and human infrastructures on a broad scale was seen as necessary for the Santomean economy to resist the oil-induced assault. As noted earlier, the Plan of Action, like Sachs's advisory project more broadly, was aimed to capture the world in all its socio-natural, technical and economic complexity in order to render it amenable for intervention. Arguably, however, at its heart remained a market-driven rationale. Any suggested "improvement" of the respective sectors—from education, to health to agriculture—serves the maximization of (people's) productivity and economic growth.[23]

In important ways, the work of Sachs's team also involved a concerted effort to manage and curtail the hopes and expectations supposedly triggered by the announcement of oil. As I noted earlier, for resource curse theorists, hope is a problem. Hopes for effortless wealth and development are deemed problematic because they are assumed to lead to increased demands on government, potentially causing instability and conflict. But just as hope generates demands, it can also lead to paralysis (cf. Crapanzano 2004, 114). Oil wealth, flowing seemingly without effort, is generally considered unproductive and associated with notions of passivity and inaction. Numerous times during my stay in STP I heard the formulaic phrase, uttered by Santomeans and expats alike, that "Santomeans do not work." In this view, the notorious rent seeking, so prominent in the resource curse literature, becomes a version of that more general Santomean worry that a people, alleged to be historically work shy, may stop working altogether and simply wait for the oil to arrive.[24]

A survey conducted by the Columbia team of people's hopes and expectations regarding oil had the effect of turning these negative externalities into something measurable and calculable. Engaging the negative externalities of hopes and expectations in this way is expected to afford a potential cure to the curse. The questionnaire asked people to choose, for example, how they would like to see the oil revenues to be spent (on education, infrastructure, health, etc.), whether a part of the money should be saved, and about levels of corruption and changes in quality of life. That is, certain kinds of hopes were permitted, but others not. Within the framework of the statistical results tables that emerged from this, irrational hopes could be made tangible, measurable, and predictable—something that could be inspected, studied, and dealt with.

Taking up Janet Roitman's analysis of the historic formation of "fiscal subjects" in the Chad Basin (Roitman 2005), the survey of people's expectations was also part of a process of constituting particular economic subjects ready to lead and sustain STP's nascent oil economy. In this context, economic subjects act upon, and are acted upon by, shifting economic practices and significations, such as those of good and bad oil wealth, distinctions of public office and private benefit, and changing relations of citizens and state expressed in the publication of accounts. The survey was perceived to have been a successful intervention in that it noticeably transformed respondents' preferences and re-constituted people as the responsible future citizens of an oil-rich country, less "self-interested" and more attuned to "an enlightened view of the common good" (Sandbu 2004; see also Humphreys, Sachs, and Stieglitz 2007). As one of the participating economists[25] declared, the survey and National Forum were an unprecedented "grand exercise in deliberative democracy" and "a real national, political exercise" (rather than a presumably "artificial" academic experiment). Despite certain flaws,[26] it is claimed to have demonstrated the effects of deliberation, and not simply of information, on people's opinions.

The efforts to implement transparency affected not only the level of government institutions but also constituted subtle attempts to change and "ethicalize" people's thought and behavior. Surveys of people's expectations and public deliberations—similar to governance capacity building programs and other "technical" assistance—are technologies (of self and other) through which specific types of ethical conduct and rational economic thought and practice, modes of calculation and estimation, and so on, are gradually folded into institutional and individual lives.[27] Quite clearly, economic theory has moved beyond the focus on changing institutions and governance arrangements, and now seeks to change local "regimes of living," that is, to engage with the finer normative, political and technical details of individual lives in response to the anticipated yet uncertain massive alterations in the constitution of the national economy (cf. Collier and Lakoff 2005, 31).

## III

If you ask a Santomean, he or she would readily tell you that hope (and greed) is not triggered by oil alone. I was reminded of this by a conversation with a Santomean friend I call José, one day quite early on in my fieldwork. It pushed me toward thinking about the ways that oil fits into local narratives of economic development in ways that take into account the historical context in which natural resource extraction occurs. José was born in STP, but spent much of his young adult life studying engineering and social policy in Portugal and the UK. Now in his early thirties, he worked in a business consultancy agency in São Tomé city. Like many of his peers who had been educated overseas and returned to the islands, he was ambivalent about the prospect of oil, noting the opportunities that it would bring to the country and to his personal life, but also the notorious volatility of the Santomean government and endemic politicking which might stand in their way.

On this day, José and I had been talking about the analytical framework of my research: Santomeans' hopes and expectations. Hope and expectation, he interrupted me, are two different things. "Africa is a continent of hope," he said emphatically. There is always hope, whereas for expectations there must be a trigger. And in STP, José continued to explain, expectations have been triggered regularly: before people expected oil and oil money, they expected the IMF and aid money. Looking at it this way, it becomes possible to identify a series of imaginaries that have determined how the past, present and future of the islands' development have been construed. They include colonial dreams of the exploitation of fertile lands through slave and contract labor on vast plantations inserted into the dense forest. Later, suggested another Santomean friend, there was the euphoria of socialism and hopes surrounding independence in 1975. Later again, people were gripped by the visions of democratic change (*mudança*) and economic "opening" (*abertura*) with the country's inclusion into a free market capitalist world.

The *abertura* brought dreams of a different kind, namely large-scale projects that could bring development to the islands and prosperity to the individuals involved. Since the 1990s, there have been regular announcements of public–private partnerships and business ventures signed between government and one or another foreign investor to build Free Trade Zones, tourism projects, or deepwater sea ports, which are propagated as bringing jobs, money and further investments. One can easily get the impression, as Nicholas Shaxson puts it, that "There was a general free-for-all in São Tomé long before oil" (Shaxson 2007, 151). Signing ceremonies for contracts with foreign investors (which are annulled with disturbing frequency and speed), and the occasional planning and partial implementation of such projects have become a vital principle structuring STP's neoliberalized temporality.

Arguably, oil is but another project of this kind, brought about by STP's varying relationship with, and insertion into, capitalist markets. Like those other projects, it holds out the possibility of development and opportunities for personal enrichment. Rumors about, and actual instances of, illicit activities surrounding such projects and the Santomean development industry, more generally, including accusations of embezzlement and corruption (cf. Seibert 2006, 196–199), are a staple in Santomean conversations and an important facet of its "cultural intimacy" (Herzfeld 1997). The actors involved may have changed over the years, but the repertoire of practices seems surprisingly similar.

*IV*

I want to conclude with the story of another Santomean friend, whose life seemed entangled in the web of new institutional practices and meanings surrounding oil's anticipated curse. I first got to know Manuel in 2007, as the assistant to STP's parliamentary committee dealing with oil, and shortly afterwards met him again as an English teacher in the local polytech college where I also gave classes during my stay. Manuel was eager to discuss the economic situation of the country with me, and drew on a variety of sources, from word of mouth to newspaper articles and Wikipedia. In addition, Manuel had participated in several of the activities conducted by NGOs and other institutions in the effort to prepare the country for oil, and had followed the political debate on oil in STP's parliament in intimate detail. Importantly, Manuel also referred to what he had learned in a petroleum business college in the Niger Delta, which he had attended as one of a group of Santomean students who had received scholarships from Nigeria to improve STP's technical capacity in this sector.

Now, however, Manuel seemed frustrated and deeply disappointed. On returning to STP after several years of training and with the expectation of landing a good job, Manuel had to discover that positions in the newly set-up National Oil Agency (supported with World Bank money and technical capacity) had been filled and that his specialized services were not needed. He sought to make sense of this experience by invoking party politics (suggesting that the National Oil Agency was dominated by individuals affiliated to what had then become the opposition party), the ineptitude of the paternalistic state (he claimed there was a failure of planning which meant that jobs for those trained abroad were in short supply); and snootiness and nepotism (according to Manuel, none of the directors of the agency had the specialized training he had but were unwilling to yield their influential positions). Perhaps they are hiding something, he mused. Despite this, and despite the increasingly disillusioning reporting on STP's oil in 2007 and 2008,[28] Manuel was convinced that one day very soon oil would come and that, if handled properly, the country could and would benefit from it.

Manuel's biography is not typical of the majority of Santomeans. However, I believe it illustrates well how the notion that oil may be a curse rather than a blessing has re-configured individual lives in the small island state. In section II, I noted the technologies that exist to induce particular forms of economic subjectivity and ethical comportment under the specter of the resource curse. People like Manuel (and others working as technicians and administrators in the institutions guarding STP's oil) were expected to embody and enact ideas of the resource curse and its potential prevention, however imperfectly. Although situated right at the heart of the new arena of meaning making in the economy of expectations, in important ways, Manuel also lacked actual power and influence and thus remained in a position of relative marginality to it.

When I returned to STP in May 2009 after an 8-month absence, Manuel had a new job. He had been hired as an administrator in the newly established Office of Records and Public Information (*Gabinete de Registo de Informações Públicas*, GRIP) responsible for the collection and publication of all the available information on oil in STP. He was still working as a teacher on the side but now earned a much better salary, had an office with his own computer, a printer, and a door to close behind him. He was busy planning a new project that would measure people's expectations on oil and work against the resource curse in STP.[29] Since its inception, GRIP has published an intermittent bulletin, established an online database, and organized a number of public seminars. More recently, however, the office's operations have suffered from the continued shortage of funds. As a result, it had to limit its operations to the provision of photocopies of official documents, and to responding to in-person and online enquiries, though efforts have been made to publicize GRIP's existence more widely through TV and radio advertisements as well as short online commercials. Manuel has successfully maintained his position as technical officer. Ironically, although the country's oil economy has failed to bring real material returns, it was the associated "transparency industry" set up in its anticipation, which had eventually provided Manuel with more secure employment.

## Conclusion

My analysis of STP's economy of expectations has sought to problematize the notion of a resource curse afflicting African oil-producer states. As my ethnographic material shows, the resource curse quickly evolved into a paradigm that has captured social scientists', policy makers' and, increasingly, the popular imagination of people living in oil-producing countries. Recent literature has been more interested in discussing the kind of social and institutional environment necessary for an oil curse to emerge, thus taking account of the many critical studies published during the last decade (e.g., Ross 2015). Yet, the political motivations and effects of the resource curse as an economic device have rarely been inspected. This paper aimed to take a first step in that direction.

I took the economic experiment conducted over the last decade in STP as a prototype of what I think is a much more general process of a re-articulation between economic theory, interdisciplinary expertise, global policy and local lives, especially in the context of natural resource developments. Sachs's project can claim some inventiveness for itself as an elaborate effort to create conditions that might prevent a curse; but its capacity to do so now threatens to dissipate. The reasons for this include, not least, the continuing absence of oil in STP, which perhaps makes the situation appear less acute and has resulted in the drying up of funds for civil society activities, with the exception of the EITI perhaps. Institutionally, Sachs's advisory project may not have had the impact it was hoped to have. Yet ideologically, it nonetheless played an important role in framing STP's emergent oil economy. It did so by setting the terms for debate, suggesting measures for a cure and offering a self-consciously holistic program to remake the small island state into an example.

At the same time, I argued that the idea of the resource curse was not introduced into STP overnight with the arrival of Sachs's advisory team. Rather, its introduction in the archipelago was more diffuse and dispersed across a range of institutions and agencies that have appropriated its discourse and are pivotal in bringing it to life. These included, as we have seen, the World Bank's own governance and capacity building program, the civil society building measures implemented by NGOs such as IA, and the steps taken by STP to be accepted as a member of the EITI. Importantly, these are the occasions where the resource curse has been performed most explicitly. But the notoriety of oil as a uniquely cursed substance is also echoed in the media, in publicly available documents, reports, newspapers, and books, and by people (both Santomeans and international personnel as well as friends and relatives in nearby Angola or Gabon).

Just as there is no "pure" resource curse, there are also no "pure" enactments of it. The resource curse is introduced in a multiplicity of forms, adjusted to the specific agendas of institutions and individuals, and articulates with existing conditions in often unpredictable ways. One of these is the thesis of the curse, an economic tool, as it is defined and debated in the academic literature. Importantly, as suggested in this chapter, it is unlikely to have originated there. The curse has now taken an intriguing new life in the form of policy recommendations, awareness-raising campaigns and technical programs, implemented by global institutions, national governments and non-governmental organizations. It has enabled global and local actors to summon new futures that pivot on notions of natural resource wealth as a potential peril as much as a bearer of hope. As people encounter the curse in different guises, they compare and contrast it with familiar ideas and instances of illicit wealth, appropriations of state property, or simply seemingly self-perpetuating patterns of social inequality. Seen in this light, the curse is less the invention of economic theorists or a possible doomed future, than a continuation of business-as-usual under slightly altered rules.

## Notes

1  Portions of this chapter are reproduced by permission from my article "Cursed resources, or articulations of economic theory in the Gulf of Guinea" published in *Economy and Society 40* (3): 345–372.
2  All names, except those of public figures, have been changed.
3  I conducted approximately 12 months of fieldwork in STP between 2007 and 2009, involving transnational and local government and regulatory institutions, industry and civil society representatives, and ordinary citizens.
4  Debates have centered, for example, on how best to conceptualize the relationship between "the economic" and "the social" (or "cultural"). For Michel Callon, economics performs its worlds, while for Daniel Miller this process is one of virtualism: the projection of abstractions onto the world, often with little success (Carrier and Miller 1998).

5  I have explored the speculative practices associated with STP's oil in a more recent article (Weszkalnys 2015).
6  According to the Santomean National Institute for Statistics, the population was 151,912 in 2006, but the number is likely to be much higher. For example, the CIA World Fact Book suggests 210,000 in July 2009.
7  The Gulf of Guinea's increasing geopolitical significance in the US American search for alternative oil supplies is indicated by the recent installation of AFRICOM, a permanent US military command in the region (McFate 2008; Omeje 2008, 8; Soares de Oliveira 2007).
8  I was able to attend some of the initial meetings of this committee in 2007 and 2008.
9  C. Hellinger, personal communication, September 2007.
10 Two licensing rounds were held for the JDZ in 2003 and 2005. Among the competitors and eventual winners were multinationals such as Chevron Texaco, Addax, and Sinopec. In 2006, one of these, Chevron Texaco, drilled an exploration well in Block 1 of the JDZ but deemed its findings not commercially viable. From 2009 to 2010, several test drills in Blocks 2, 3, and 4 were conducted by Addax Petroleum and by the Chinese Sinopec Corporation. In July 2010, Chevron Texaco's interests in Block 1 were acquired by the multinational Total. However, the French multinational later abandoned its interests following additional test drills. A further licensing round for STP's EEZ was held in 2010, and exploration activities are currently under way. For more extensive discussions of the details of these oil deals see, e.g., Seibert (2006, 2008, 2013), Soares de Oliveira (2007), Shaxson (2007).
11 See Weszkalnys (2016).
12 See Weszkalnys (2014).
13 The Open Society Institute has been particularly active in the realm of democracy and civil society building, sponsoring both research and advocacy. See www.soros.org/.
14 See also the project website: www.earthinstitute.columbia.edu/cgsd/stp/index.html (last accessed April 20, 2010).
15 This type of "useful" economics frequently reaches out to interdisciplinary knowledge. See also the analysis of the Earth Institute's work in the context of a broader contemporary effort to institutionalize interdisciplinary research (Weszkalnys and Barry 2012).
16 This is not the place to discuss Sachs's more general approach to global economics. He has come under critique for his more recent visions for development, which emphasize the role of geography and environment and, arguably, neglect institutional, political and historical factors that may lead to under-development (interviews with Earth Institute researchers, May 2006).
17 There has been a curious doubling up of efforts, characteristic of the Santomean economy of expectations. In addition to the law, STP signed the so-called Abuja Declaration, which expresses a will to transparency in the JDZ. Moreover, aside from the Columbia team's legal advice, the World Bank had also hired a team of legal experts to develop an oil law. Similarly, the oil revenue management law that was eventually adopted sees for the set up of an Oversight Commission, rather similar to those initiated by the EITI, which has also been implemented in STP. In 2008, there was some heated debate regarding the unnecessary duplication of institutions and the respective value of a "global" or "local" initiative.
18 See Ross (1999, 301–302) for earlier critiques of conventional development economics.
19 My development of the concept of articulation has much benefited from discussion at the workshop "Translations of travelling legal, organizational, and techno-scientific models in African contexts" at the VAD conference, Mainz 2010, organized by Richard Rottenburg, Johanna Mugler, and Andrea Behrends.
20 *New York Times*, July 2, 2007.
21 This coup d'état, as a previous coup in 1995, did not involve any bloodshed. STP has had a relatively peaceful history, at least since the height of the colonial period,

culminating in an uprising against a forced-labor policy and the massacre of an estimated 1000 local people in 1953 (Seibert 2006, 87). Significantly, unlike other Portuguese colonies, STP did not have a war of liberation; its decolonization was essentially directed from the Lisbon diaspora (Seibert 2006, 118).

22 www.earth.columbia.edu/cgsd/stp/documents/bulletin_english_3.1.pdf (last accessed September 30, 2010).

23 The Plan of Action was never officially presented, and only partially pursued by the Columbia team. The Columbia advisory project has dwindled away, operating on increasingly reduced funding, although ties between team and country remain. The Santomean government, including the erstwhile enthusiastic president, has taken courses of action partly quite different from those recommended to bring sustainable development by Sachs's action plan.

24 There is a very powerful discourse on the supposed "indolence" of Santomeans, which transcends racial and social class boundaries. This supposed aversion to work is often considered to be grounded in STP's history of slavery and forced labor, where the rejection of strenuous, physical labor became a sign of freedom and social status (Seibert 2006). The discourse on "indolence" has indeed proved persistent throughout STP's colonial as well as post-colonial socialist and neoliberal periods, and is now re-interpreted within the framework offered by resource curse theories. See Weszkalnys (2016) for a longer discussion.

25 Sandbu's interdisciplinary background—in philosophy, politics, and economics with a focus on behavioral economics—partly reflects the spirit of the project.

26 Humphreys, Sachs, and Stieglitz's analysis of the Forum demonstrates considerable leader effects, that is, the influence that the age and gender of the discussion leaders had on the final results of group deliberation (Humphreys, Sachs, and Stieglitz 2007). This is a potentially significant finding, throwing a different light on the benefits of participatory activities. However, the authors brush aside such more fundamental questions, suggesting that a better design would rule out the problem. Another point of contention in the context of the National Forum was the payment of local representatives who would participate in the final discussion. As documented on the project's website, here, the exercise of disinterested citizenship mixed, unexpectedly perhaps, with economic self-interest.

27 Another example is the attempt to induce entrepreneurial behavior in Santomeans and to encourage "local content." In 2008, for example, a business forum was held bringing together Santomean and Nigerian businessmen to discuss entrepreneurial activities for the JDZ, that is, possibilities to exploit the resources in the Zone apart from oil. More generally, the notion that local service providers and businesses need to be strengthened to stand the challenge of the emergent oil economy and cater to its needs, and that local business capacities need to be enhanced and civil society built—so prevalent in contemporary STP—further underscores this point (see also Weszkalnys 2016).

28 In 2006, Chevron—one of the operators in a major block in the JDZ—had done some test drills that were found to be "not commercially viable." But many people I talked to, including Manuel, thought that Chevron was lying, deliberately delaying production to increase its profits.

29 He was advised, however, that this project was somewhat outside the remit of his position.

# References

Auty, Richard. 1993. *Sustaining Development in Mineral Economies: The Resource Curse Thesis*. London: Routledge.

Barry, Andrew. 2004. "Ethical Capitalism." In *Global Governmentality*, edited by Wendy Larner and William Walters, 196–211. London: Sage.

Barry, Andrew, and Slater, Don. 2002. "Introduction: The Technological Economy." *Economy and Society 31* (2): 175–193.

Behrends, Andrea. 2008. "Fighting for Oil When There Is No Oil Yet: The Darfur-Chad Border." *Focaal 52*: 39–56.

Bell, Joseph C., and Teresa Maurea Faria. 2007. "Critical Issues for a Revenue Management Law." In *Escaping the Resource Curse*, edited by M. Humphreys, J. Sachs and J. Stiglitz, 286–321. New York: Columbia University Press.

Brunnschweiler, Christa N., and Erwin H. Bulte. 2008. "Linking Natural Resources to Slow Growth and More Conflict." *Science 320* (5876): 616–617.

Çalişkan, Koray, and Michel Callon. 2009. "Economization, Part 1: Shifting Attention from the Economy towards Processes of Economization." *Economy and Society 38* (3): 369–398.

Çalişkan, Koray, and Michel Callon. 2010. "Economization, Part 2: A Research Programme for the Study of Markets." *Economy and Society 39* (1): 1–32.

Callon, Michel. 1998. "The Embeddedness of Economic Markets in Economics." In *The Laws of the Markets*, edited by Michel Callon, 1–57. Oxford: Blackwell.

Carrier, James G., and Daniel Miller, eds. 1998. *Virtualism: A New Political Economy*. Oxford and New York: Berg.

Collier, Paul, and Anke Hoeffler. 2000. "Greed and Grievance in Civil War." World Bank Policy Research Working Paper 2355. Washington, DC: World Bank.

Collier, Steven J., and Andrew Lakoff. 2005. "Regimes of Living." In *Global Assemblages: Technology, Politics, and Ethics as Anthropological Problems*, Steven J. Collier and Aihwa Ong, 22–39. Malden: Blackwell.

Crapanzano, Vincent. 2004. *Imaginative Horizons: An Essay in Literary-Philosophical Anthropology*. Chicago and London: University of Chicago Press.

Earth Institute at Columbia University (Earth Institute). 2008. *The Investor's Guide to São Tomé and Príncipe*. New York: Earth Institute at Columbia University. *www.afrst.illinois.edu/outreach/business/profiles/sao/. . ./saotomeinvestorguide.pdf.*

Ferguson, James. 2006. *Global Shadows: Africa in the Neo-Liberal World Order*. Durham and London: Duke University Press.

Ferguson, James. 2015. *Give a Man a Fish: Reflections on the New Politics of Distribution*. Durham and London: Duke University Press.

Fine, Ben. 1998. "The Triumph of Economics; or, 'Rationality' Can Be Dangerous to Your Reasoning." In *Virtualism: A New Political Economy*, edited by James G. Carrier and Daniel Miller, 49–73. Oxford and New York: Berg.

Frynas, Jędrzej George, Geoffrey Wood, and Ricardo M. S. Soares de Oliveira. 2003. "Business and Politics in São Tomé e Príncipe: From Cocoa Monoculture to Petro-state." *African Affairs* 102 (406): 51–80.

Gelb, Alan and Associates. 1988. *Oil Windfalls: Blessing or Curse?* New York: Oxford University Press.

Guala, Francesco. 2007. "How to Do Things with Experimental Economics." In *Do Economists Make Markets? On the Performativity of Economics*, Donald MacKenzie, Fabian Muniesa, and Lucia Siu, 128–162. Princeton and Oxford: Princeton University Press.

Haber, Stephen, and Victor Menaldo. 2007. "Do Natural Resources Fuel Author-itarianism? A Reappraisal of the Resource Curse." Working paper 351. Stanford: Stanford Center for International Development.

Herzfeld, Michael. 1997. *Cultural Intimacy: Social Poetics in the Nation-State*. New York and London: Routledge.

Holmes, Douglas R., and George E. Marcus. 2005. "Cultures of Expertise and the Management of Globalization: Toward the Re-functioning of Ethnography." In *Global Assemblages: Technology, Politics, and Ethics as Anthropological Problems*, edited by Aihwa Ong and Steven J. Collier, 393–416. Malden, and Oxford: Blackwell.

Humphreys, Macartan, Jeffrey D. Sachs, and Joseph E. Stiglitz, eds. 2007. *Escaping the Resource Curse*. New York: Columbia University Press.

Karl, Terry Lynn. 1997. *The Paradox of Plenty: Oil Booms and Petro States*. Berkeley: University of California Press.

Klein, Naomi. 2007. *The Shock Doctrine*. New York: Penguin Books.

Krastev, Ivan. 2004. *Shifting Obsessions: Three Essays on the Politics of Anticorruption*. Budapest and New York: Central European University Press.

Latour, Bruno. 1987. *Science in Action: How to Follow Scientists and Engineers through Society*. Cambridge, MA: Harvard University Press.

Leite, Carlos and Jens Weidmann. 1999. "Does Mother Nature Corrupt? Natural Resources, Corruption, and Economic Growth." Working paper, WP/99/85. Washington, DC: International Monetary Fund.

Leonard, Lori. 2016. *Life in the Time of Oil: A Pipeline and Poverty in Chad*. Bloomington: Indiana University Press.

MacKenzie, Donald A., Fabian Muniesa, and Lucia Siu, eds. 2007. *Do Economists Make Markets? On the Performativity of Economics*. Princeton: Princeton University Press.

Massey, Simon, and Roy May. 2005. "Dallas to Doba: Oil and Chad, External Controls and Internal Politics." *Journal of Contemporary African Studies 23* (2): 253–276.

Massey, Simon, and Roy May. 2010. "The Collapse of the 'Doba Model' and its Impacts on Chadian Politics and People." Paper presented at the African Studies Association of the UK, Oxford, September.

McFate, Sean. 2008. "Briefing: US Africa Command: Next Step or Next Stumble." *African Affairs 107* (426): 111–120.

Mitchell, Timothy. 2005. "The Work of Economics: How a Discipline Makes its World." *European Journal of Sociology 46* (2): 297–320.

Obeng-Odoom, Franklin. 2015. "Global Political Economy and Frontier Economies in Africa: Implications from the Oil and Gas Industry in Ghana." *Energy Research & Social Science 10*: 41–56. doi: 10.1016/j.erss.2015.06.009.

Omeje, Kenneth C. 2008. "Extractive Economies and Conflicts in the Global South: Re-engaging Rentier Theory and Politics." In *Extractive Economies and Conflicts in the Global South: Multi-Regional Perspectives on Rentier Politics*, edited by Kenneth C. Omeje, 1–25. Aldershot: Ashgate.

Peel, Michael. 2009. *A Swamp Full of Dollars: Pipelines and Paramilitaries at Nigeria's Oil Frontier*. London: I.B. Tauris.

Pegg, Scott. 2005. "Can Policy Intervention Beat the Resource Curse? Evidence from the Chad-Cameroon Pipeline Project." *African Affairs 105* (418): 1–25.

Peluso, Nancy Lee, and Watts, Michael, eds. 2001. *Violent Environments*. Ithaca and London: Cornell University Press.

Reno, William. 1999. *Warlord Politics and African States*. Boulder: Lynne Rienner.

Reyna, Stephen P. 2007. "The Traveling Model that Would Not Travel: Oil, Empire, and Patrimonialism in Contemporary Chad." *Social Analysis 51* (3): 78–102.

Reyna, Stephen P., and Andrea Behrends. 2008. "The Crazy Curse and Crude Domination: Towards an Anthropology of Oil." *Focaal 52*: 3–17.

Rigterink, Anouk S. 2010. "The Wrong Suspect: An Enquiry into the Endogeneity of Natural Resource Measures to Civil War." Paper presented at the CSAE Conference 2010: Economic Development in Africa, March 21–23.

Roitman, Janet. 2005. *Fiscal Disobedience: An Anthropology of Economic Regulation in Central Africa.* Princeton: Princeton University Press.

Ross, Michael L. 1999. "The Political Economy of the Resource Curse." *World Politics* 51 (2): 297–322.

Ross, Michael L. 2015. "What Have We Learned about the Resource Curse?" *Annual Review of Political Science 18* (1): 239–259.

Rosser, Andrew. 2006a. "The Political Economy of the Resource Curse: A Literature Review." IDS Working Paper. Sussex, UK: Institute of Development Studies, University of Sussex.

Rosser, Andrew. 2006b. "Escaping the Resource Curse." *New Political Economy 11* (4): 557–570.

Sachs, Jeffrey D. 2000. "Notes on a New Sociology of Economic Development." In *Culture Matters*, edited by Lawrence E. Harrison and Samuel P. Huntington, 29–43. New York: Basic Books.

Sachs, Jeffrey D. 2005a. "Ending Poverty: How Universities Can Help." Paper presented at the 2005 Aspen Symposium, September 26 www.earthinstitute.columbia. edu/sitefiles/file/about/director/documents/EndingPovertyHowUniversitiesCanHelp-AspenSymposium-Sept262005_editedSAF_.pdf.

Sachs, Jeffrey D. 2005b. *The End of Poverty: Economic Possibilities for Our Time.* New York: Penguin Books.

Sachs, Jeffrey D., William Masters, Vijay Modi, Awash Teklehaimanot, and the Columbia University Advisory Project in São Tomé and Príncipe. 2005. *Towards a Consensus Plan of Action for São Tomé and Príncipe.* New York City: CGSD, The Earth Institute, Columbia University.

Sachs, Jeffrey D., and Andrew M. Warner. 1995. *Natural Resource Abundance and Economic Growth.* Cambridge: Harvard Institute for International Development.

Sachs, Jeffrey D., and Andrew M. Warner. 2001. "The Curse of Natural Resources." *European Economic Review 45* (4–6): 827–838.

Sala-i-Martin, Xavier, and Arvind Subramanian. 2003. "Addressing the Natural Resource Curse: An Illustration from Nigeria." NBER Working Paper, No. W9804. Cambridge, MA: National Bureau of Economic Research.

Sandbu, Martin E. 2004. "Does Deliberation Matter? A Study of Deliberative Democracy in São Tomé e Príncipe's National Forum," September. http://citeseerx. ist.psu.edu/viewdoc/download?doi=10.1.1.691.6463&rep=rep1&type=pdf.

Satre, Lowell J. 2005. *Chocolate on Trial: Slavery, Politics and the Ethics of Business.* Athens: Ohio University Press.

Seibert, Gerhard. 2003. "Coup d'état in São Tomé e Príncipe: Domestic Causes, the Role of Oil and Former 'Buffalo' Battalion Soldiers." ISS paper 81. Pretoria: Institute for Security Studies.

Seibert, Gerhard. 2006 [1999]. *Comrades, Clients and Cousins: Colonialism, Socialism and Democratization in São Tomé e Príncipe.* Leiden and Boston: Brill.

Seibert, Gerhard. 2008. "São Tomé and Príncipe: The Troubles of Oil in an Aid-dependent Micro-state." In *Extractive Economies and Conflicts in the Global South: Multi-Regional Perspectives on Rentier Politics*, edited by Kenneth C. Omeje, 119–134. Aldershot: Ashgate.

Seibert, Gerhard. 2010. "Wo bleibt das Öl. In São Tomé and Príncipe Frustriert der ausbleibende Ölboom Lokale Politiker und ausländische Philantropen." In *Edition Le Monde Diplomatique 05: Afrika—Stolz und Vorurteile*, 70–75.
Seibert, Gerhard. 2013. "São Tomé and Príncipe: The End of the Oil Dream?" IPRIS Viewpoints 134 (September). Lisbon: Portuguese Institute for International Relations and Security (IPRIS).
Shaxson, Nicholas. 2007. *Poisoned Wells: The Dirty Politics of African Oil*. New York and Basingstoke: Palgrave Macmillan.
Soares de Oliveira, Ricardo. 2007. *Oil and Politics in the Gulf of Guinea*. London: Hurst & Co.
Tsing, Anna Lowenhaupt. 2000. "Inside the Economy of Appearances." *Public Culture* 12 (1): 115–144.
Vicente, Pedro C. 2010. "Does oil corrupt? Evidence from a Natural Experiment." *Journal of Development Economics 92* (1): 28–38.
Vines, Alex. 2010a. *An Uncertain Future: Oil Contracts and Stalled Reforms in São Tomé e Príncipe*. Human Rights Watch Report.
Vines, Alex. 2010b. "Locking in Success: The Rise of China-Sonangol." Paper presented at the African Studies Association of the UK, Oxford, September.
Watts, Michael. 2001. "Petro-Violence: Community, Extraction, and Political Ecology of a Mythic Commodity." In *Violent Environments*, edited by Nancy Lee Peluso and Michael Watts, 189–212. Ithaca and London: Cornell University Press.
Watts, Michael. 2004. "Resource Curse: Governmentality, Oil, and Power in the Niger Delta, Nigeria." *Geopolitics 9* (1): 50–80.
Weszkalnys, Gisa. 2008. "Hope and Oil: Expectations in São Tomé e Príncipe." *Review of African Political Economy 35* (3): 473–482.
Weszkalnys, Gisa. 2009. "The Curse of Oil in the Gulf of Guinea: A View from São Tomé e Príncipe." *African Affairs 108* (433): 679–689.
Weszkalnys, Gisa. 2014. "Anticipating Oil: The Temporal Politics of a Disaster Yet To Come." *Sociological Review 62* (S1): 211–235.
Weszkalnys, Gisa. 2015. "Geology, Potentiality, Speculation: On the Indeterminacy of 'First Oil.' *Cultural Anthropology 30* (4): 611–639.
Weszkalnys, Gisa. 2016. "A Doubtful Hope: Resource Affect in a Future Oil Economy." *Journal of the Royal Anthropological Institute 22* (S1): 127–146.
Weszkalnys, Gisa, and Barry, Andrew. 2013. "Multiple Environments: Accountability, Integration and Ontology. In *Interdisciplinarity: Reconfigurations of the Social and Natural Sciences*, edited by Andrew Barry and Georgina Born, 178–208. London and New York: Routledge.
Winner, Langdon. 1993. "Upon Opening the Black Box and Finding it Empty: Social Constructivism and the Philosophy of Technology." *Science, Technology, and Human Values 18* (3): 362–378.

# 8 Illness, compensation, and claims for justice

## Lessons from the Choropampa mercury spill

*Fabiana Li*

On June 2, 2000, a truck traveling from the Yanacocha Gold Mine in Northern Peru spilled 151 kilograms (334 pounds) of liquid mercury along a 27-kilometer stretch of highway traversing the towns of Choropampa, Magdalena, and San José. The truck belonged to the company RANSA, a subcontractor of *Minera Yanacocha* (the Yanacocha Mining Company), which operates the country's largest gold mine. Mercury is an inconvenient by-product of the gold mining process; to sell it, Minera Yanacocha must transport it more than 600 kilometers from the mine in the Northern province of Cajamarca to the capital city of Lima. On the day of the spill, children and other residents took notice of the shiny substance on the ground and tried to collect it with bare hands, spoons, and even with their mouths. In the days following the accident, more than 750 people sought medical attention for symptoms of acute mercury exposure, which include chest pain, nausea, cough, fever, and dermatitis,[1] resulting from inhalation and contact with the skin. Minera Yanacocha's public statements affirm that the company dealt with the consequences of the spill, compensated those affected, and ensured that the soil and water were free of mercury contamination[2]. The company considers Choropampa a closed case, but the disaster lives on in the experiences of local residents and in the public imaginary.

In this chapter, I revisit the mercury spill and its aftermath to consider what this incident can tell us about environmental governance at a time of increased investment in resource extraction and rapid expansion of mining activity in Latin America. In particular, I am interested in examining how the strategies of the corporation, the actions of state institutions, and residents' contested claims about toxicity and compensation shaped local responses to an environmental disaster. The Choropampa spill coincided with the development of large-scale mining projects in Peru that brought conflict and controversy to many parts of the country in the early 2000s. Spurred by neoliberal reforms that created favorable investment conditions for multinational corporations, mining expansion also brought with it new risks associated with open-pit mining, cyanide-leaching technologies, and toxic byproducts like mercury. The Choropampa case offers a stark example of how the lives of people in marginalized communities are put at risk to sustain our current economic order: in this case, a model of development based on the aggressive expansion of resource extraction into areas that did not previously depend on extractive industries.

Studies of resource extraction have emphasized the uneven benefits derived from oil, gas, and mining, which often operate as enclaves, seemingly detatched from their surrounding societies (Ferguson 2006). In contrast to this image of isolation, the Choropampa spill shows that the consequences of mining activity extend far beyond communities immediately surrounding the Yanacocha mine. Yet outside the area defined by the company as the mine's direct "area of influence," people do not benefit from the limited employment and social programs offered by today's modern mines. The mercury spill is a story about those spaces "on the side of the road" (Stewart 1996), excluded from the "progress" and opportunities promised by extractive industries yet transformed by their operations. In Choropampa, the daily flow of mine vehicles passing through the town emphasize people's exclusion from the purported benefits of mining while serving as a tangible reminder of the risks that accompany extractive activity.

In spite of recent conflicts related to Minera Yanacocha's operations (Li 2015), the mining company seeks to present a positive corporate image of environmental responsibility and good relationships with local communities. Company-sponsored environmental monitoring programs, industry standards of performance, and other voluntary initiatives broadly defined as "Corporate Social Responsibility" (CSR) enable mining companies to incorporate some ethical concerns into their operations while evading others. The Choropampa spill brings to light how these new ethical formations are "defined, authorized and performed" (Dolan and Rajak 2016) as communities and corporations evaluate, measure, and contest the risks of mining activity. The events in Choropampa also elucidate how forms of knowledge are mobilized (or marginalized) as different actors variously responded to and sought recognition of the effects of the spill.

The Choropampa case raises questions about what constitutes adequate knowledge, what counts as evidence, and how the two are intertwined. In the literature, the concept of "popular epidemiology" has been used to describe conflicts between professional and lay ways of knowing about environmental health risks (P. Brown 1992, 1997; Novotny 1994). In some cases of toxic exposure, experiential knowledge is valued over (and sometimes precedes) expert assessments and official scientific awareness. As the events in Choropampa also show, the reluctance of corporate and government actors to study, treat, and prevent environmentally induced diseases can lead to a lack of trust in traditional science and the state, which is exacerbated by secrecy or the withholding of information (P. Brown 1992).

The ways in which authorities and the public deal with various types of environmental disasters can also reflect different "civic epistemologies" (Jasanoff 2008): shared understandings of what credible claims should look like, and how they ought to be articulated, represented, and defended in public domains. In some contexts technical information and "trust in numbers" (Porter 1995) occupy a more privileged place. Porter argues that this is a characteristic of modern societies, but "trust in numbers" takes a distinctive form in different national contexts, based on the particular historical conditions and political-economic institutions that influence the adoption of quantification and cost-benefit valuation (Fourcade 2011). In other contexts, the authenticity of experience may count for more than the assurances of powerful experts (Jasanoff 2008). This results in two incompatible ideas of accountability: on the one hand, public reason, articulated through law, science, or economics; and on the other, public morality, with demands for social justice and the assumption of responsibility by the state.

This chapter explores how different notions of accountability, evidence, and justice contributed to the various layers of knowledge and epistemic failure that made up the Choropampa disaster. I use the term "epistemic failure" to examine the limitations or the absence of legitimized ways of knowing to explain the effects of toxicity on the population. Epistemic failures make it difficult to demand change or seek compensation for environmental harm, and can lead to a breakdown in relations between community, state, and corporate actors. I begin by showing how the events following the spill were produced by the neglect and irresponsibility of company engineers, lawyers, doctors, government officials, NGO workers, and others. The effects of the accident were compounded by the company's attempt to deal with, conceal, compensate, or minimize the effects of the mercury. These actions were enabled by state institutions, not due to their absence—as critics sometimes charge–but by their public avowal of and private deference to the company's decisions, codes of conduct, and standards of "responsibility." The company relied on environmental studies and compensation schemes that were used to demonstrate corporate accountability, yet these efforts had the effect of concealing the human dimensions of the disaster.

In the second part of the chapter, I contrast the dominant legal, medical, corporate, and state narratives with those of local people, whose experiences and claims against the company challenged the assumption that expert knowledge is accurate and conclusive. I describe how bodies became a political battleground, as people turned to the law to demand justice but were not always able to support their claims with quantitative measures, such as test results or other scientific evidence. What people sought was a collective response that recognized the tragedy as a responsibility of the state and society as a whole. Instead, the company's seemingly arbitrary compensation schemes exacerbated divisions within the community and weakened collective action, illustrating that the individualization of injury and compensation is both a central aspect of neoliberalism and a key to its success.

## Negligence and complicity

In a short online video titled "Choropampa: 15 Years without Answers" (Gran Angular 2015), released on the 15th anniversary of the mercury spill, a

Choropampa resident declares: "For Yanacocha, this is a closed case, but not for us; on the contrary, because of the symptoms and aches every day, it's an open wound." Residents of Choropampa and the Cajamarca region frequently referred to the spill as an "open wound," and this image has galvanized protestors and become a powerful symbol for resistance to resource extraction. The Choropampa spill played an important role in shaping mining-related activism, and it has also served as a case study (and cautionary tale) in the mining industry's development of Corporate Social Responsibility programs (Welker 2013). My interest in Choropampa stems from research on mining conflicts involving the Yanacocha mine, a joint venture between the US-based Newmont Gold Company, the Peruvian company Buenaventura, and the International Finance Corporation, the financial arm of the World Bank (Li 2015). During my fieldwork in Cajamarca in 2005 and 2006 and in follow-up visits in 2010 and 2012, I examined people's experience of the spill through visits to Choropampa, which is 48 kilometers from the city of Cajamarca, and interviews with community leaders, villagers, and NGOs advocating for victims. Additionally, I draw on films, newspaper coverage, NGO reports, and company materials produced from the time of the spill until its 15th anniversary in 2015. Although the spill affected two other communities, in this paper I refer primarily to Choropampa, where most of the victims resided.

Choropampa is a town of approximately 600 inhabitants living in the urban core whose economic activities revolve primarily around commerce, including a weekly market. It is traversed by the principal highway leading from the city of Cajamarca to the coast, the route taken by numerous trucks carrying hazardous materials to and from the mine. Based on socio-economic indicators and vital statistics, Choropampa and the surrounding region are characterized by high rates of poverty, low levels of education, and limited access to basic health services (Defensoría del Pueblo 2001), all of which contributed to how people experienced the spill and its aftermath.

The Choropampa spill shattered the image of "safe" responsible mining that the Yanacocha Mining Company had been trying to create since it began operating in 1994. Yanacocha was one of the first "mega-mining" projects that followed the implementation of neoliberal economic reforms during the government of Alberto Fujimori, part of a larger trend in the Latin American region that generated a number of popular struggles in response to extractive activity (see Bebbington and Bury 2013; Szablowski 2007). Neoliberal reforms facilitated foreign investment and brought with them expectations of economic growth and employment opportunities. However, the specialized labor force required in modern mines did not match local expectations for jobs, development projects, and a better standard of living. Instead, mining brought conflicts over land and water, and fears about the unknown dangers of chemical mining. Until the spill, most Cajamarca residents had not known that mercury was a product of gold mining. This lack of knowledge contributed to the intense distrust and conflict that continues to characterize the relationship between the company and communities surrounding the mine.

The spill took place in the afternoon of June 2, 2000, as a transport truck operated by Minera Yanacocha's subcontractor RANSA passed through the towns of San José, Choropampa and Magdalena, located approximately 250 kilometers from the mine. Children playing on the streets were the first to see the small silver droplets on the ground, which eventually captured the attention of other townspeople, who began to handle the mercury without recognizing the risks involved. Adults believed it to be a valuable substance; some knew it as *azogue*, a substance believed to have curative properties.[3] By noon the following day, company and local government representatives arrived to inspect the site of the spill. Hoping to recuperate the mercury that was lost, mine personnel offered 100 soles (around US$35) for each kilogram of mercury recovered. Minera Yanacocha put RANSA personnel in charge of buying back the mercury (CAO 2000), establishing a buy-back system in a small store that continued for up to a week after the spill.

Residents were hired for clean-up operations, and were paid to help recuperate the mercury with brooms, buckets, and sacks, without any protective gear. José, who was in his early twenties and lived in Choropampa, was one of the people hired for the clean-up work. He told me most workers were not aware of the risks involved, because in those first few days, people had not yet started to get sick from the mercury. The pay seemed good, and he did not develop any symptoms until much later, when he wondered if the headaches he got after reading for long periods might be related to the mercury, and if it might have affected his memory. Most of all, people resented the company's way of handling the spill. "They acted like they owned the town," said José, complaining about how Yanacocha representatives treated people like liars and did not own up to their responsibility for the accident.

In addition to the company, other actors—and the knowledge they created—shaped the response to the spill. These included government representatives from the Ministry of Health and the Ministry of Energy and Mines, and the Compliance Advisor Ombudsman (CAO), an organization created by the International Finance Corporation (IFC) and Multilateral Investment Guarantee Agency to respond to complaints from communities affected by projects with IFC involvement, such as the Yanacocha mine. In a report written on behalf of the non-profit organization ECOVIDA, Arana (2000) summarizes a number of irregularities with the initial response to the spill. On the part of the state, Arana finds that the Ministry of Health acted late and inefficiently. The report also notes that the doctor who was responsible for the medical attention provided from the start was Director of the Cajamarca Regional Hospital, and at the same time, a medical advisor to Yanacocha (and therefore, one of its employees). The doctor initially maintained that the symptoms reported were caused by a viral infection, while other medical professionals diagnosed people with leukemia. Many patients were not given their blood and urine test results or questioned their validity. The Regional Director of Education refused to suspend classes, even as children fell ill and clean-up operations continued—

this was to avoid projecting an image of a contaminated town, purportedly for the benefit of Minera Yanacocha.

The Ministry of Energy and Mines did not play a significant role in the immediate response to the spill, and did not have the technical expertise or resources to deal with environmental disasters. A few weeks after the spill, the Ministry fined Yanacocha for the amount of 600 UITs (US$500,000), which went directly to the Ministry and did not benefit the affected population. Maria Luisa Cuculiza of the Ministry for Women arrived in Choropampa to tell residents not to engage in legal actions against the company or trust any lawyers, because President Fujimori would be their best lawyer (Cabellos and Boyd 2002). The response of state authorities reveals not only a lack of knowledge about the dangers of mercury, but deliberate negligence and complicity in downplaying the effects of the spill. While state institutions are often faulted for their lack of leadership and engagement in mining conflicts, what we can also see in this case is their support for the company and efforts to influence public opinion. At the time, it was imperative for the government to defend the mining industry, which was being touted as an increasingly significant player in the country's economic development. Their interventions contributed to an erosion of trust in the state and the company, making people skeptical about subsequent information provided by the company, state actors, and medical experts.

Negligence also characterized the actions of the companies involved. RANSA's contract with Minera Yanacocha required the transport company to comply with the safety regulations and environmental standards, but the absence or lack of enforcement of state laws leaves companies like Minera Yanacocha with the task of establishing their own standards of performance and ensuring the compliance of their subcontractors. In fact, RANSA's contract with Minera Yanacocha offered few details about the handling of dangerous materials. These oversights could have been noticed through closer supervision and regular inspections to ensure that RANSA was prepared to respond to a toxic spill, but Minera Yanacocha did not conduct these inspections. The subcontracting of labor common in today's transnational mining operations also brings with it a deferral of responsibility, as Minera Yanacocha's slow response to the spill illustrates.

In a study carried out by the CAO, observers faulted Yanacocha for not having adequate policies and procedures for handling and transporting hazardous waste or an appropriate emergency response plan for spills occurring outside its property. The company was also found to have underestimated the dangers posed by the mercury on human health and the environment, and to have under-reported the amount of mercury spilled (CAO 2000). Indeed, one of the first uncertainties surrounding the spill had to do with the amount of mercury that was recovered, and the amount that remained in the environment. Over time, this figure acquired more precision, based on the company's own calculations: Minera Yanacocha reported that out of 11.1 liters spilled (or 151 kilograms),[4] 92.8 percent were recovered by clean-up crews, 3.4 percent was temperature-evaporated and 3.8 percent was not locatable (Minera Yanacocha 2013).

The company repaved the highway and monitored soil and air conditions in the areas of the spill. It also monitored air quality in homes, and in Choropampa, 67 homes were identified as having some level of mercury contamination. Homes with mercury readings higher than the recommended levels underwent a clean-up that included the removal of dirt and construction materials. According to Luis Quequejana, a chemist working for the Ministry of Health at the time of the spill, the government failed to evacuate people from contaminated homes, and deferred to Minera Yanacocha's instructions for the clean-up (Boyd 2003). Many houses had dirt floors and adobe walls that would have made it difficult to completely eliminate traces of mercury.

Amidst the lack of organization and information over those first few days, Minera Yanacocha played up its collaboration with authorities and medical specialists, including a team of international doctors and toxicology specialists tasked with assessing the consequences of the spill. These medical experts were brought in to Choropampa to reassure residents that the effects of mercury are not permanent. Michael Kosnett, a toxicologist from the University of Denver hired by Yanacocha, explained to the media that once an individual is removed from the source of mercury exposure, the symptoms they experienced would be gone in a matter of weeks (Panorama 2009). In an interview, one resident told me he recalled being told the mercury would be naturally eliminated by the body, just like beer. According to a company report, the mercury had been completely removed after the clean-up operations, and specialists agreed that the spill posed "no risks of future health problems as a result of the accident" (Minera Yanacocha 2013).

Minera Yanacocha implemented a series of environmental mitigation measures, focusing on water, soil, and sediment in the areas where mercury was known to have spilled and establishing soil and water monitoring points. It performed 2 years of Environmental Risk Assessment Studies, carried out by the international consulting firm Shepherd Miller (based in the United States), which included sample analyses of water, plant and animal life. The company concluded that following remediation efforts, there was no longer any danger of contamination: "The results of the samples of plants, insects, animals, and soils demonstrate that concentrations of mercury found are far below the normal limits of any living being or soils in other zones. The same is true for the aquatic environment" (Minera Yanacocha 2004). The company decided when to stop the studies and monitoring, based on its own interests and assessment of the problems.

While many questions about the spill remained unanswered, Minera Yanacocha's environmental studies and monitoring results gave scientific legitimacy to the company's claims. These methods of risk assessment create an image of control and careful management even in cases of high uncertainty. The same is true of preventive measures, such as health and safety standards and procedures, put in place by companies, as well as predictive methods, like modeling, that assess the possible risks of mining. Technological optimism can lead scientists and engineers to focus more attention on well-defined, short-term risks rather than indeterminate, long-term ones (Jasanoff 2003). In this case, the company's Environmental Impact Assessments (which are presented to government authorities before the start of any mining project) failed to adequately prepare the company for a toxic spill, yet the Ministry of Energy and Mines approved the project. After the accident, the company's environmental studies shifted away from the health dimensions of the spill, including the multiple variables that might have determined people's levels of exposure, exacerbated their symptoms, or led to long-term problems. While environmental studies and risk assessments project objectivity, completeness, and scientific rigor, they also demarcate the parameters of what is considered "relevant" information, ignoring the risks and uncertainties that might extend beyond the limits established in the studies (Goldman 2005). In the Choropampa case, environmental studies of soil, water, and animals excluded the human dimensions of the accident. Framing the spill in environmental terms sidesteps the messiness of the human experience of the disaster, including the political, economic, and ethical claims made by those affected.

Bracketing out uncertainty and complexity has the effect of making the science wielded by experts seem "objective" and therefore worthy of being treated seriously, while other forms of knowledge are labeled "subjective" and "anecdotal" (K. Brown 2013, 312). In dealing with the spill's impacts on human health, the company emphasized its reliance on scientific studies and toxicity testing, projecting an image of impartiality and fairness in compensating those affected by the spill. From the point of view of local residents, however, science, in the hands of doctors, scientists, and authority figures, failed them time and time again. They remember being misinformed by medical experts about the effects of mercury poisoning, and receiving conflicting information by state and company representatives. Medical science did not cure the symptoms that people continue to attribute to the spill, and some people claim that the medicine they were given to fight the effects of mercury poisoning, such as penicillamine, may have produced long-term side effects, as will be discussed below. In the next section of the chapter, I show how the company relied on test results for mercury poisoning to establish eligibility for compensation. Missing paperwork, unreliable test results, and unfair compensation practices contributed to people's mistrust and the feeling that their illnesses and grievances were made invisible by the absence of quantitative evidence.

## Illness, compensation, and the law

The Sanchez family owned a small restaurant on Choropampa's dusty main road, the obligatory route for trucks and other vehicles traveling between the coast and the city of Cajamarca. Sonia, a single mother of three daughters and a son, prided herself on her hard work and resourcefulness, qualities that her children emulated. The mercury spill upended their lives, not only because of their exposure to the mercury, but because their restaurant became a cafeteria for workers and out-of-town visitors dealing with clean-up efforts after the spill. The spill made the close-knit Sanchez family into victims and employees of the mining company, putting them in a precarious situation: they aroused the ire of the townsfolk as they briefly benefitted from the boon to their restaurant, but were left with financial difficulties as business died down and their hopes for compensation failed to materialize. The family's plight encapsulates the profound consequences of the spill on people's health, economic situation, and hopes for the future.

Sonia's youngest daughter, Angela, was only 12 years old when she handled the mercury while she was out playing with a group of children. Five years later, she told me about the events of that day:

> We were racing toy cars, when suddenly we saw something shiny . . . then we made a game of who could find the most [mercury]. We collected it, because we thought it was pretty. With our hands, with paper, with a spoon, because you couldn't collect it just like that. We put it in bottles. I remember that day I collected it in a bottle, but when people started saying "it's bad, it's toxic," it occurred to me to throw it in the *pozo ciego* (septic tank).

After the spill, representatives of Minera Yanacocha offered Sonia 4 months of work providing meals to people involved in the clean-up efforts. Her children helped with the business, and they fed 300 people a day working in the aftermath of the spill, including Yanacocha personnel, security, clean-up crews, and even local teachers. During this period there was a boomtown feel to Choropampa: everyone had work, and people wanted to take advantage of the opportunity, knowing it would not last. While they had work, the Sanchez family felt they could not complain or they would have lost their contracts. The company purchased new kitchen equipment and expanded the restaurant to fit more customers. Sonia felt betrayed by her conflicting loyalties. She had to put up with the anger of townspeople who felt her family had sold out and who even tried to vandalize the restaurant.

Some days after the spill, Angela developed symptoms of mercury poisoning.

> I had chills, fever. My head hurt. I tried to ignore it. It always hurt but I ignored it. Sometimes I didn't eat so I could make it to the restaurant to help my mom. One of those times, I was helping with the dishes, and I felt dizzy. My head hurt. I fainted.

Angela was taken to the hospital and was among those who were given medication to counter the effects of the mercury, and developed anemia and gastritis as side effects. Sonia's two youngest daughters were both hospitalized for mercury poisoning, but they quickly reached a settlement in the initial rounds of compensation. Angela's sister, Patricia, wanted to study, and used the money to help pay for her university education. Years later, Sonia and her daughters regretted reaching a deal with the company too quickly, but her work contract with Minera Yanacocha compelled her to accept their terms and give up any possibility of taking legal action in the future.

After the clean-up work ended and people were left feeling abandoned, the problems and the protest actions intensified. The company's efforts to deal with the aftermath of the spill were not always voluntary, and were not necessarily demanded by the state, but were the result of direct pressure from residents who organized marches and roadblocks to demand justice. Choropampa residents spoke heroically about the first protests, which were held in February of 2001. Sonia's eldest daughter, Dani, was in her twenties with two children of her own. When I met her in 2005, she held out hope that she would receive compensation so that she could pay for their studies. What she and other towns-people wanted most, however, was for the company to provide all residents with long-term health insurance. Dani was among the women who participated in the roadblocks and marches even though they had never been involved in political activism. She described her motivation for joining the protests thus:

> We were demanding what is just, our health insurance, because it was given to some, who had their test results [showing traces of mercury in blood and urine]. But those of us who didn't have them, we didn't get anything. No medicine, nothing . . . that's what the protest was about.

Townspeople had the support of members of the *Rondas Campesinas* (peasant patrol groups) who came down from communities in the surrounding mountains. The protests attracted attention in Cajamarca and beyond, as the spill became a symbol of the dangers associated with mining—even the modern methods of extraction that transnational companies branded as clean and responsible mining.

The protests and activism surrounding the spill forced the company to negotiate individual and community compensation. In Choropampa, the community compensation program included the construction of a new school and sports complex, improvements to existing schools, the expansion of the medical clinic, and remodeling of the town square. The company also invested in improving water and drainage systems and paving roads. Total expenditures for these community projects were 2,590,000 Peruvian Soles, (or approximately US$800,000), while an additional 660,000 soles was spent in the town of San Juan, and another 377,000 soles in Magadelena (Minera Yanacocha 2013). Additionally, an insurance policy was set up for 1173 residents of the affected areas to treat problems related to mercury exposure over a period of 5 years.

Alongside the protests and public pressure, people turned to the law to demand justice, and hoped that the transnational nature of the company might enable them to demand accountability in the United States. As with other accidents and environmental disasters, lawyers rushed to Choropampa ready to make the most of the situation. They initially promised large sums of money, but soon lawyers encouraged those affected by the spill to sign extra-judicial settlements. Out of need and desperation, people who had been exposed to the mercury agreed to small amounts of compensation in spite of the potential long-term effects that the spill might have on their health, and did not insist on long-term health insurance.

The compensation money was not enough for people to improve their economic situation, and the amounts seemed to be based on randomly determined criteria and quantitative measures, on a case-by-case basis. Adding to this perceived unfairness were the inconsistencies with lab results and mercury testing, which were used to determine the amounts given in the form of compensation. Minera Yanacocha (2013) reported that around 4000 people underwent medical exams overseen by the Ministry of Health, of which 935 had mercury levels above the recommended levels of 20 micrograms per liter (some had up to 135 μg/L). The first to be compensated received between US$600 and US$6000 (Boyd 2003), but by signing the settlement, they lost the right to claim any further damages from the company. Using lab results to determine how much people should be given in compensation might have been seen by company representatives as way to be objective, but the company reached agreements and out-of-court settlements with different people at various times. The inconsistency of this approach created confusion and a sense of injustice.

As in other cases of compensating for environmental damages, disagreement, and conflict emerge around questions of commensurability (Li 2011), especially when trying to attribute monetary value to intangible things. As Fourcade (2011) also argues, how various cases of environmental pollution (such as oil spills) are handled depends on the institutional acceptability of money as a yardstick of value. Whether or not the natural environment can be monetized depends on organizations—courts, public agencies, hospitals, and corporations—that influence the process of valuation and render it legitimate. Based on her study, Fourcade argues that social institutions operate differently in the United States and France, noting that in the US, there is greater social acceptance for monetary compensation for injury. The US legal system is used as a compensatory instrument, but what is also notable is that compensatory processes have increasingly taken a rationalized form, "relying on specialized expertise to produce economic value" (Fourcade 2011, 1734).

In the Choropampa case, we see how the mining company established monetary equivalences outside of the legal system in out-of-court settlements, a compensatory process that was partially accepted by Peruvian institutions which lacked a precedent on which they could draw and local residents whose disadvantaged position left them unable to negotiate. Also evident is an incomplete process of rationalization, which inconsistently demanded blood tests and others "facts" from outside experts to determine compensation amounts. However, the government institutions responsible were ill equipped to handle mercury testing and the treatment of sick patients; people lacked confidence in these institutions, and did not see them as impartial arbiters advocating on their behalf. Part of the problem had to do with the test results themselves. Some people had tested negative in spite of exposure and illness after the spill; people also wondered why family members who had been equally exposed showed different test results. Others were suspicious about why their test results were high at first, then quickly went down in the next days. Some claimed that they were not shown their results when the tests were first taken, and that the results from testing done by DIGESA (General Health Division) had gone missing. All of this contributed to rumors and the suspicion that doctors, scientists, and government officials had been paid off to hide the real magnitude of the disaster.

Sonia's experience exemplified the confusion and lack of trust generated by the laboratory tests. She recounted how one early morning, as she was working at the restaurant, she came down with a fever started feeling pain in her lungs. She was taken to the local health clinic, but her test results came out negative: "Everything in the blood, in the urine, everything disappeared. How strange, don't you think? Right then and there, I realized it was an injustice. They were mocking us." She could not comprehend why the test results did not show the presence of mercury, and asked:

> If we are all human beings, why did some come out with mercury poisoning, and us, we have nothing? . . . Just there it should be evident to anyone, because we are all human beings . . . It's as if we were to make a comparison with cooking some eggs. Once the water is boiling, we put them all in, but only half get cooked and the others stay raw. Can that be?

Sonia's questions expressed her frustration at the devaluation or denial of her plight, and her wish for a resolution that recognized the townspeople's common experience, their dignity, and their common humanity. She also desired some form of reparation for her family's ills, but these ills were not quantifiable in medical terms—indeed, they were not solely medical, and not only related to the mercury spill.

Some months after the spill, the California-based firm Engstrom, Lipscomb & Lack arrived in Choropampa and promised to sue Newmont, Yanacocha's main shareholder, in the United States. The firm had gained notoriety for representing residents of Hinkley and Kettleman, California, in a class action lawsuit surrounding groundwater contamination that was made famous by the Hollywood film *Erin Brockovich*. Some 1000 people affected by the mercury spill signed on to the mass tort, a separate process from a Peruvian legal suit that was also launched. The plaintiffs in the US lawsuit included Sonia's oldest daughter, Dani, and her two children. They were the only ones in the family who did not reach a settlement with the company in the initial rounds of compensation. Dani's two children had the lab paperwork to show elevated mercury levels after the spill. For those who had them, these test results became prized possessions—their only chance at getting a higher sum of money, either in or out of court. Like other victims of environmental injustice (see Auyero and Swistun 2009), Dani and her family were caught between the present and the future, dealing with their daily hardships while holding out hope for a settlement or a successful lawsuit. The promise of compensation, no matter how uncertain, was one of the only ways to improve their lives, and perhaps, to leave Choropampa in search of other opportunities.

A couple of months before we spoke in August 2005, the company started offering people who were part of the lawsuits 2000 soles (US$620) as compensation for those with zero or very low mercury levels at the time of the spill. They signed an agreement at a meeting with all the plaintiffs, but most of those with high test results held out. Dani accepted the 2000 soles because she needed it for medications and she had lost hope of getting a better deal. "There's no proof," she said, referring to her negative test results, which showed a value of 0.0 (she was referring to micrograms per liter, the standard measurement for exposure). A normal mercury level is less than 10 µg/L (micrograms/liter) and less than 20µg/L in urine, and the absence of any trace of mercury was suspicious to her, indicating the possibility of fraud.

The lab results had become people's only bargaining tool. By accepting this compensation, she gave up the right to make any more claims, but her two children remained part of the US lawsuit. Since she did not agree to the offer they made for them, she pondered the question of how much money was the just amount. She said that ideally, what she wanted was lifetime health insurance for her children. Those who signed on with the lawyers found themselves with their hands tied, having been left with a case that seemed to be going nowhere and no communication with lawyers in Lima and California who were supposed to be defending them. Dani told me of the difficult economic situation in Choropampa. There was no money, no jobs, and the family restaurant was not doing well. After the spill, vehicle traffic through the town declined, and many people moved away. It was obvious that the US$100,000 per family that they had sued for with the US law firm was their one hope for economic compensation. Even if it was a long shot, it was still their only chance to improve their lives. But their economic hardships made it more and more difficult to hold out, and as time went by, many were getting ready to accept an out-of-court settlement.

The lack of unity among the townspeople did not allow for collective bargaining—each person was on his or her own, and the company dealt with each case individually and inconsistently. Part of the weakening of political organizing had to do with the compensation—some chose to hold out, while others accepted the company's terms, creating divisions within the community. There were rumors about who had received money and how much. Juana Martinez, the president of the Defense Front and one of the key players in the initial protests, was later believed to have received compensation. That's why she was not as vocal, people said, and why she suddenly had a new truck. These kinds of rumors were rampant and divisive, ultimately limiting political activism and benefitting the company.

Among the rumors and conflicting information, conducting fieldwork in Choropampa brought with it its own uncertainties. As people told me about their symptoms and recounted the events of June 2000, many things remained unclear and no amount of testing could ascertain what the effects of the mercury spill had been. People's symptoms could not be definitively tied to the mercury, nor could incidents of animals born deformed, increased rates of miscarriages, or people falling ill. Their testimonies, told to NGO workers and journalists, were a way to create the evidence that they lacked, and a way to make visible an injustice that was otherwise fading from public view. Outsiders accused Choropampa residents of consciously manipulating the situation in the hope of receiving compensation money. What could I make of people's claims, opinions, and perceptions when there was legal action in progress, when there was a lawsuit at stake? On the other hand, were the "facts" produced by Minera Yanacocha and others any more valid, if they were used to downplay the effects of an environmental disaster with undeniable consequences for the population?

The story of Choropampa is one in which rumors, fiction, and facts collide, mix, and become indistinguishably blurred. My goal is not to disentangle them. Instead, I want to suggest that the best way to analyze this and other environmental disasters is to consider what contested and uncertain knowledges can tell us about justice, expertise, and corporate accountability. Adriana Petryna (2002) writes that in "discerning the 'true' causes of their subjects' suffering, researchers themselves have inadvertently reified categories of authentic and inauthentic suffering, marginalizing those who happen to fall in the latter category" (12). In her own study, she seeks to avoid pigeonholing people affected by the Chernobyl disaster as suffering from either "hard" biologically induced symptoms or "soft" psychological ones. A similar approach might be necessary in the context of Choropampa and other environmental disasters, leaving aside the question of causality and evidence to focus on the experiences of those whose lives have been adversely affected.

Even if the mercury was no longer visible, people lived with the feeling that they were already contaminated. Choropampa was a sick town, and everyone I talked to complained of some illness or symptoms that they recognized might or might not be related to the mercury. But there was always that possibility, accompanied by the resentment of not having the money to make themselves well again. Of course, most people also recognized that no amount of money would make them well, or make up for the grave injury to the community. However, in such situations, money serves as a signifier of authenticity, helping to legitimate their suffering and their claims of physical and psychological damages resulting from the spill. The same symptoms came up repeatedly in conversations: headaches, bone aches, stomach problems, numbness in the limbs, and memory loss. Dani said half-jokingly: "The spill has left us forgetful, nervous, and choleric."

The doctor at the Community Health Clinic and an assistant were in charge of dealing with people covered by Minera Yanacocha's five-year health insurance plan. The insurance expired in June 2005, but was extended for another 6 months, and then again for a year because of pressure from towns-people. When patients came to the clinic, Dr. Torres' first task was determining if their symptoms could be related to mercury exposure. When I interviewed him in 2005, he was seeing 60 to 70 people a month who qualified for treatment (this did not mean that they were suffering from the effects of mercury, only that they qualified for treatment according to the health insurance guidelines).

The problems Dr. Torres saw were primarily neurological and dermatological: headaches and dizziness, and skin rashes and hives. There were people who tried to take advantage of the insurance, he said, and wanted to be treated for respiratory problems or lipomas, which had nothing to do with the mercury. However, it was possible that the symptoms—especially the rashes—could indeed be an aftereffect of the spill, although he said a study would need to be done to confirm this. Another medical problem he mentioned related to the side effects of the medication, including penicillamine, which was given to counteract the effects of mercury poisoning. He said it was possible that people were given a high dosage, or that people self-medicated and took more than was recommended. The medication is known to have side effects such as gastritis and digestive problems. The doctor had been compiling medical histories for people's files used for legal purposes, and noticed inconsistencies in the mercury tests that were done after the spill. This coincided with testimonies of people who claimed that the results had been manipulated. Patients at the clinic seemed to be keeping close tabs on their medical visits, and asked for photocopies of their medical reports that were then sent to their lawyers.

Another staff member at the clinic was Maribel, who had been working there for about 2 years at the time of my visit. When I asked her if the effects of the mercury were still being felt, she said that this was more of a psychological issue and there were now very few cases of people going to the clinic with problems, at least compared to the number of people they used to see. She said the cases may or may not be related to the mercury, there was no way to tell. The problems people complained about were too general, like joint and bone pain, headaches and stomach problems. They could only give them *calmantes* (pain killers) so although people wanted something that would deal with their health problems, all the clinic could do was provide temporary relief. People may have stopped going to the clinic knowing that they could not get any real treatment other than ibuprofen and vitamins that were handed out for free, and many resorted to other treatments like natural medicines.

When I asked Maribel about the rashes some people complained of, she told me they might be measles or chicken pox—that they were starting a campaign because there was something of an epidemic, and that the rashes may have been misdiagnosed. This seems like an implausible explanation, since these rashes were recurrent. She also told me that in many ways, Choropampa was what it was because of Minera Yanacocha—referring to the development projects and money spent on paving roads and fixing other infrastructure. Otherwise, Choropampa would be just another "pueblito," an impoverished small town. Her position as a healthcare worker and a non-local interacting with residents through their visits to the clinic gave her a specific role in the community. Her positive view of Minera Yanacocha likely influenced how she saw the patients, and she seemed convinced that the effects of the spill were mostly gone.

In spite of the skepticism often shown by those around them, people's sick bodies enabled them to externalize what could not be expressed in other ways. Their medical symptoms were a way to call attention to the grave injustice caused by a powerful multinational whose mining operations did not benefit them directly. Except for the community development projects, and in spite of the compensation some people received, they did not feel that the company's actions had significantly improved their difficult economic situation. People's demands were based on moral rather than scientific or technical arguments. The irony in this is that the body (and the perpetuation of people's sickness) became the only way in which disempowered groups could legitimate their claims and seek some form of justice. Illness and suffering can thus become cultural resources through which people stake claims for social equity (Petryna 2002), citizenship, and the right to healthcare and a clean environment.

## Conclusion

After years of waiting, the lawyers from Engstrom, Lipscomb & Lack met with their clients in January 2008 in a Cajamarca hotel. Each person was called upon to sign a document stating the amount that they would each be awarded, and to give up the right to any further claims or legal actions against the company. Everything was to remain strictly confidential. Only afterwards did people find out the wide disparities in the compensation figures: anywhere from $1000 to $75,000, and a couple of the settlements reportedly paid out more than $100,000 (Luna Amancio, 2008). No one found out the total amount awarded in the collective settlement for the mercury spill, and they were not informed about the details of the private arbitration process because they had already signed a power of attorney allowing the lawyers to reach an agreement on behalf of the affected residents.

For Minera Yanacocha, the legal case has long been closed, yet for those affected, the injustice remains. The mercury spill in Choropampa shows how the multiple vulnerabilities of the population—signified by poverty, unemployment, and limited access to basic services such as healthcare—magnified the consequences of the disaster and the inadequacy of the response to it. Examining the mercury spill calls attention to the company's strategies to avoid taking responsibility, and the complicity of experts, corporate actors, and state authorities that supported the company's decisions in response to the spill. Their words and actions obfuscated the effects of the accident, and contributed to the contradictory knowledge, confusion, and epistemic failures that characterized the disaster and its aftermath. Furthermore, the company's efforts to monitor and assess the environmental conditions after the spill helped create an image of corporate accountability and scientific rationality while making other things invisible—including the uncertainties relating to long-term impacts of the spill, and the company's responsibility for the health of the affected communities.

In the aftermath of the spill, laboratory test results and mercury levels came to define what counted as evidence and helped to establish a monetary equivalent in compensation claims. The reliance on quantitative results and (at times dubious) scientific evidence delegitimized the experiences of local people and their first-hand accounts of the disaster. One of the problems was the lack of information, as in the case of lost or unreliable lab records, but the available test results (which showed negative or low values for mercury poisoning) invalidated the experience of illness of those affected, adding to their sense of powerlessness and magnifying the inequalities that already existed.

The individualization of injury and compensation contributed to the fragmentation of the community and collective action. The company's handling of the spill obliged people to accept the notion that their symptoms and experiences could be monetized, creating a climate of competition and distrust. However, alongside claims for compensation, people's claims and protests reflect a desire for collective recognition and acknowledgment that the company's liability extends into the future, beyond the immediate needs of individuals. These different conceptions of justice, time, responsibility, and evidence made the body into a political terrain that will continue to be marked by controversy and contested narratives.

## Acknowledgments

I would like to thank Lori Leonard and Siba N. Grovogui for convening and hosting the "Governing Extractive Industries" workshop at Cornell, where this material was presented, as well as the workshop participants for the fruitful discussions that emerged. I also thank Jessica O'Reilly, Rima Praspaliauskiene, Brenda Austin-Smith, Roisin Cossar, Shawna Ferris, Krista Johnston, Susan Prentice, and Vanessa Warne for reading and helping me improve upon earlier drafts of this chapter; Carolina Meneses for research assistance; and Pablo Lapegna for his constructive suggestions for revisions.

## Notes

1  Chronic mercury exposure can bring on symptoms such as tremors, emotional changes (e.g., irritability, nervousness), insomnia, and neuromuscular changes.
2  Mercury released into the environment can enter lakes and streams, affecting water quality and wildlife.
3  Some people use *azogue* to cure "susto" (literally, "fright," a condition seen to affect babies in particular).
4  The company used liters as the unit of measurement to report the amount of mercury spilled, while other sources used kilograms.

## References

Arana, Marco A. 2000. *Informe de la Verdad sobre el Desastre Ambiental en Choropampa.* Cajamarca, Peru: ECOVIDA.

Auyero, Javier, and Debora Alejandra Swistun. 2009. *Flammable: Environmental Suffering in an Argentine Shantytown*. Oxford: Oxford University Press.

Bebbington, Anthony, and Jeffrey Bury, eds. 2013. *Subterranean Struggles: New Dynamics of Mining, Oil, and Gas in Latin America*. Austin: University of Texas Press.

Boyd, Stephanie. 2003. *Choropampa: The Price of Gold*, edited by Stephanie Boyd. Booklet. Lima, Peru: Guarango Film and Video. January–February: 1–24. http://icarusfilms.com/guide/cho.pdf.

Brown, Kate. 2013. *Plutopia: Nuclear Families, Atomic Cities, and the Great Soviet and American Plutonium Disasters*. Oxford: Oxford University Press.

Brown, Phil. 1992. "Popular Epidemiology and Toxic Waste Contamination: Lay and Professional Ways of Knowing." *Journal of Health and Social Behavior* 33 (3): 267–281.

Brown, Phil. 1997. "Popular Epidemiology Revisited." *Current Sociology* 45 (3): 137–156.

Cabellos, Ernesto, and Stephanie Boyd. 2002. *Choropampa: The Price of Gold*. Guarango Film and Video. Brooklyn: First Run/Icarus Films.

Compliance Advisor Ombudsman (CAO). 2000. *Investigación del Derrame de Mercurio del 2 de Junio del 2000 en las Cercanías de San Juan, Choropampa, y Magdalena, Peru*. Lima, Peru: CAO.

Defensoría del Pueblo. 2001. *Informe Defensorial No. 62: El Caso del Derrame de Mercurio que Afectó a las Localidades de San Sebastián de Choropampa, Magdalena y San Juan, en la Provincia de Cajamarca*. Lima: Defensoría del Pueblo.

Dolan, Catherine, and Dinah Rajak. 2016. *The Anthropology of Corporate Social Responsibility*. New York: Berghahn.

Ferguson, James. 2006. *Global Shadows: Africa in the Neoliberal World Order*. Durham, NC: Duke University Press.

Fourcade, Marion. 2011. "Cents and Sensibility: Economic Valuation and the Nature of 'Nature.'" *American Journal of Sociology* 116 (6): 1721–1777. doi: 10.1086/659640.

Goldman, Michael. 2005. *Imperial Nature: The World Bank and Struggles for Social Justice in the Age of Globalization*. New Haven: Yale University Press.

Gran Angular. 2015. Choropampa: 15 Años sin Respuestas. www.youtube.com/watch?v=V7bz2O2lils.

Jasanoff, Sheila. 2003. "Technologies of Humility: Citizen Participation in Governing Science." *Minerva* 41 (3): 223–244. doi:10.1023/A:1025557512320.

Jasanoff, Sheila. 2008. "Bhopal's Trials of Knowledge and Ignorance." *New England Law Review* 42 (4): 679–692. doi: 10.1086/518194.

Li, Fabiana. 2011. "Engineering Responsibility: Environmental Mitigation and the Limits of Commensuration in a Chilean Mining Project." *Focaal: Journal of Global and Historical Anthropology*. 60: 61–73.

Li, Fabiana. 2015. *Unearthing Conflict: Corporate Mining, Activism, and Expertise in Peru*. Durham, NC: Duke University Press.

Luna Amancio, Nelly. 2008. "Tras 11 Años del Derrame de Mercurio, Síntomas Persisten en Choropampa." *El Comercio*. May 22. http://elcomercio.pe/planeta/761410/noticia-11-anos-derrame-mercurio-sintomas-persistenchoropampa.

Minera Yanacocha. 2004. *Balance Social 2003*. Cajamarca: Minera Yanacocha.

Minera Yanacocha. 2013. *Fact Sheet: Choropampa*. English Translation. January. Cajamarca: Minera Yanacocha.

Novotny, Patrick. 1994. "Popular Epidemiology and the Struggle for Community Health: Alternative Perspectives from the Environmental Justice Movement." *Capitalism Nature Socialism* 5 (2): 29–42.

Panorama. 2009. *Choropampa: El Precio del Oro Ocho Años después. Part II.* September 25. www.youtube.com/watch?v=VXPcgGfubIE.

Petryna, Adriana. 2002. *Life Exposed: Biological Citizens after Chernobyl.* Princeton: Princeton University Press.

Porter, Theodore M. 1995. *Trust in Numbers: The Pursuit of Objectivity in Science and Public Life.* Princeton: Princeton University Press.

Stewart, Kathleen. 1996. *A Space on the Side of the Road: Cultural Poetics in an "Other" America.* Princeton: Princeton University Press.

Szablowski, David. 2007. *Transnational Law and Local Struggles: Mining, Communities, and the World Bank.* Portland: Hart.

Welker, Marina. 2014. *Enacting the Corporation: An American Firm in Post-Authoritarian Indonesia.* Berkeley: University of California Press.

# 9 Wars of words

## Experts, oil, and environmental governance in Chad

*Lori Leonard*

On September 29, 2011, Oxfam America hosted a lunch-time report launch and panel discussion at the Carnegie Endowment for International Peace in downtown Washington, DC. At the center of the event was *Watching the Watchdogs*, Oxfam America's 93-page report on the workings of three expert panels that monitored the implementation of social and environmental safeguard policies in oil and gas projects in Chad, Georgia, and Peru. Expert monitors are a new addition to the suite of environmental governance mechanisms attached to extractive industry projects and are fast becoming *de rigueur* in high-risk and high-profile projects. Global oil and mining companies and the international financial institutions that back them have embraced expert monitors in response to calls for transparency and public accountability in the extractive sector.

The emphasis on enhancing transparency and accountability in oil and gas projects stems, in part, from the "curse" of oil, or the much cited empirical observation that countries with natural resources, and particularly oil, tend to have slower rates of growth and more conflict, political instability, and problems with financial mismanagement than countries with fewer such resources (Auty 1993; Karl 1997). Efforts to promote transparency in the extractive sector have mostly focused on money flows, as in the case of the Extractive Industries Transparency Initiative (EITI), in which governments and companies disclose payments received and made (Haufler 2010). But global oil companies and their international financial backers have also come under increasing scrutiny for the ways their modes of operating contribute to environmental degradation and social dislocation, and the oil industry has worked hard to shed its reputation as dirty and dangerous, even as a spate of recent incidents, including the Deepwater Horizon blowout in the Gulf of Mexico and a series of exploding oil trains across the American Midwest, belie the clean, modern, and technologically sophisticated image the industry has tried to project.

In *Watching the Watchdogs*, Oxfam America suggested that disclosures about the social and environmental practices of governments and global oil companies could alter their operations in ways that would improve outcomes for local communities. Expert monitors tasked with making these disclosures are expected to operate as independent and impartial witnesses. Oxfam rated the three expert panels it reviewed on criteria such as independence, transparency, stakeholder engagement, and influence and on an array of related measures imagined to make these groups more or less effective. *Watching the Watchdogs* identified expert monitors as a critical part of the expanding transparency assemblage in the extractive sector while refocusing attention on the question of how transparency and information disclosure are imagined to work. Scholars have approached this question from different angles.

Arthur Mol (2006) characterized efforts to connect the disclosure of knowledge and information with voluntary behavior modification on the part of companies that degrade or befoul the environment as "rather naïve" (498) and urged social scientists to think more critically about the constitutive and formative roles knowledge and information play in the Information Age (Castells 1996, 1997a, 1997b). Mol's arguments extend the work of Manuel Castells, particularly the distinction Castells (1996) made between an information economy and an informational economy, to the domain of environmental governance. Conventional modes of environmental governance rely on the authority of science and the capacity of nation-states to use scientific knowledge to formulate laws and regulations. Mol suggests that this mode of environmental governance persists, but that in the Information Age environmental struggles are also lifted out of specific localities and played out in the media and through flows of information and symbols (see also Mol 2008). Environmental governance increasingly takes the form of struggles over information within broad and diverse networks, where no actor has the ability to control the outcome of environmental debates. In the informational economy, information is a vital resource and the ability to access, produce, manage, control, and disseminate it is not evenly distributed but is key to productivity, influence, and competitiveness.

In contrast to this work, which is attuned to how information is produced and used, scholars of the extractive industries have focused on the performative aspects of information disclosure. Oil, gas, and mining companies are under pressure to show concern for local communities and environments but face the problem of how to demonstrate a commitment to ethics (Dolan and Rajak 2011). One response has been to adopt standards and best practices and incorporate them into corporate social responsibility platforms. Transparency has become a watchword in these efforts, and companies have assembled people, processes, and technologies to help them manage the disclosure of information; expert monitors are only the latest addition to these corporate repertoires of openness. This work centers not so much on the content of what is shared as on what performances of transparency produce. It focuses, for example, on how transparency simultaneously reveals and conceals, and how disclosure works to politicize or de-politicize situations and to expand and contain discussion and action (Rajak 2011; Barry 2013). In short, this work describes how transparency operates as a technique of governmentality that transforms institutions, communities, and oil economies.

In this chapter I draw on these different perspectives in exploring how expert monitors and information disclosure shaped environmental governance in the Chad–Cameroon Pipeline Project. I am interested in the role experts played because of the immense differences in the capacities of actors in extractive industry projects—especially the Chad project—to mobilize information. In Chad, the project operated as a joint venture between ExxonMobil, which represented a consortium of global oil companies in Chad, the governments of Chad and Cameroon, and the World Bank. Expert monitors have been positioned as "the new transparency powerbrokers" (Mol 2010, 139) and are expected to act as third-party mediators of competing informational claims. Yet how expert mediation works in oil and gas projects, where the gaps in the capacities of actors to produce, manage, and disseminate information are particularly acute, is an open question.

I am also interested in the role of expert monitors in environmental governance because of how they have been championed as useful adjuncts in the ethics project of big oil. The World Bank joined the venture in Chad to enact a model of resource governance and to transform Chad into an oil economy that would avert the resource curse. Expert monitors were expected to play critical roles: to exert normative pressure on oil companies through information disclosure and build local capacity to audit future oil projects. Oxfam America formulated a checklist of "best practices" for expert monitors and concluded its evaluation by recommending that they be required for all publicly-financed extractive industry projects (Oxfam America 2011, 8). Panelists at the *Watching the Watchdogs* launch, including oil and gas industry representatives, concurred. As expert monitors become standard features of the transparency assemblage and key to oil futures, it becomes vital to ask how they operate and what their participation means for ethics, openness, and oil-as-development.

Mol wrote that in the Information Age no actor can fully control the outcomes of environmental debates, but in this chapter I argue what might be the opposite side of that coin: that it is also impossible for actors to stand outside those debates as independent, objective, or neutral parties. This means that any account of the work that experts do has to take seriously their immersion in networks and flows of information. I do this by presenting ethnographies that show some of those entanglements and complicate the linear picture of environmental governance offered in publications like *Watching the Watchdogs*. The ethnographies illustrate how experts and their disclosures shaped environmental outcomes in the project while never losing sight of experts' necessarily interdependent and entangled positionalities. I suggest that claims about experts' independence, objectivity, and neutrality should not be read as technical findings or statements of fact but treated as informational resources and analyzed for the work they do. In the Chad project, transparency had the effect of limiting and containing debates about ethics to the standards regime and drawing attention to the consortium's commitment to that regime even as it was breaching it.

## A "model" project

The model project the World Bank assembled in Chad was supposed to help the country "escape the resource curse" (Polgreen 2008) and show that an extractive industry project could function as a poverty reduction project. Efforts to mitigate the social and environmental impacts of the project were central to that mission. In fact, the pipeline project in Chad was the World Bank's most complete attempt to engineer an extractive industry project as a poverty reduction project. The ability of the Bank to create such a comprehensive model for oil-as-development reflected, in turn, Chad's standing as one of the poorest and least stable countries in the world as well as the poor quality of the oil in its Doba Basin fields and the substantial cost and complexity of building a 1070-kilometer pipeline from southern Chad across Cameroon to the port of Kribi on the Atlantic Ocean. ExxonMobil invited the participation of the World Bank—a move that reduced its own risk and exposure to negative publicity for a highly controversial project and gave the World Bank added leverage to set policy. The project was a blueprint for others because only in Chad did the World Bank have such a free hand in project design.

The World Bank described the project in Chad as a template for other extractive industry projects because of the unprecedented package of conditionalities and policy provisions it was able to write into its loan agreements. The government's use of oil revenues was governed by a revenue management law that channeled oil money into social sectors of the economy, reserved oil revenues for development initiatives in the oil-producing region, and sequestered funds in offshore accounts for future generations. The consortium of oil companies agreed to comply with an array of social and environmental standards, including the World Bank's safeguard policies and industry standards and best practices for its operations, including construction methods, waste management, compensation and resettlement, worker health and safety, and community consultation. Esso, the ExxonMobil affiliate that acted on behalf of the consortium in Chad, codified these agreements in a multi-volume Environmental Management Plan (EMP) that was vetted by the World Bank as a condition of project approval. The World Bank also financed a Petroleum Sector Management Capacity Building Project that anticipated the expansion of Chad's oil economy beyond the three fields in the Doba Basin region, and funded the development of legal and regulatory infrastructure for Chad's emerging oil economy as well as the creation of geophysical databases and the institutional capacity to negotiate and manage future oil deals.

The success of the project as a poverty reduction project was imagined to hinge on transparency—on the ability of expert monitors to track and report to the public on the implementation of these agreements and thereby hold the Chadian government, the World Bank, and ExxonMobil accountable. The World Bank created multiple monitoring bodies with overlapping duties. In Chad, the *Collège de Contrôle et de Surveillance des Ressources Pétrolières* (CCSRP) included government officials and members of civil society who tracked the use of deposits made to Chad via a London bank account. A second local monitoring body, the *Comité Nationale de Suivi et du Controle* (CTNSC), was an inter-ministerial committee charged with shadowing consortium staff in the field and monitoring their compliance with technical, social, and environmental standards on behalf of the government. The CTNSC was supposed to test air, water, soil, and dust samples alongside the consortium; develop a national oil spill response plan; deal with problems related to sanitation and the spread of HIV and AIDS; and manage forests at risk of being decimated by the influx of migrants to the oil field region (IAG July 2002). The work plans of the CCSRP and the CTNSC included training in how to monitor complex and technologically sophisticated oil projects, foreshadowing the future expansion of Chad's oil sector and Chad's emergence as an oil economy.

Other expert monitoring groups assembled by the World Bank duplicated the functions of local groups and provided training to them. The External Compliance Monitoring Group (ECMG) for the Chad pipeline project was an Italian firm that monitored the environmental impacts of the project and the consortium's compliance with its own environmental risk mitigation plan. The World Bank's Inspection Panel, which responds to complaints from people and communities adversely affected by a Bank-funded project, was also part of the multi-layered system of supervision and monitoring. The International Advisory Group (IAG)—the expert monitoring body featured in the Oxfam America report—was responsible for advising the World Bank president and the governments of Chad and Cameroon on how the project was advancing and how it could achieve its poverty reduction goals. Because it was the monitoring body with the most ambitious and encompassing mandate, it is also the monitoring body I focus on most closely in this chapter.

I observed the functioning of these expert monitoring bodies in *canton* (or county) Miandoum over the period from 2001 to 2012—from the beginning of the project until 4 years after Chad paid off its debts to the World Bank and the Bank withdrew from the project, quietly acknowledging that its experiment in resource-driven development had failed. I followed the project through the lives of 80 families in *canton* Miandoum, one of three *cantons* at the heart of the Doba Basin project and home to the Miandoum oil field. Expert monitors visited the villages of the *canton* infrequently, but residents were entangled in the consortium's ethical protocols, especially those governing the expropriation of their land. If the problem for global oil companies is how to demonstrate ethical behavior, the consortium staked its claim to ethics and corporate social responsibility by making its plans for social and environmental risk mitigation visible in Chad but also—and especially—outside of it. The Environmental Management Plan and expert monitoring bodies were central to this effort.

## Visible templates, invisible tensions

The IAG was a five-person body whose members were appointed by the World Bank on the basis of their expertise, or "eminence," in their fields (World

Bank, nd). The chair of the panel was a former prime minister of Senegal as well as a lawyer and an economist. The four additional members contributed expertise in agriculture, civil engineering, environmental policy, and cultural anthropology. The IAG operated with considerable autonomy.

The group had an open mandate and an annual budget of roughly US$600,000 (Oxfam America 2011). The IAG maintained its own secretariat in Montreal and was supposed to "independently develop its work program" (World Bank 2001, 3). The *Terms of Reference* the World Bank established for the IAG indicate that the group had "freedom to obtain information from all relevant sources" (World Bank 2001, 3) and could operate without restrictions on data collection. Members were expected to have access to reports prepared by or on behalf of the consortium, the governments of Chad and Cameroon, and the World Bank Group, and were free to meet regularly with members of these groups and other "stakeholders" (World Bank 2001, 3), including NGOs and farmers in *canton* Miandoum. The IAG set its own work schedule. In most years, the group made two trips to Chad, but there was the possibility of more frequent supervision, "depending on the IAG's assessment" (World Bank 2001, 3). Yet, despite the openness of its mandate, the IAG opted to adopt the EMP as "the enforcement document for social and environmental issues for the duration of the project" (IAG February 2004, 2) and to place the EMP at the center of its monitoring efforts.

The EMP consisted of 20 volumes and roughly 5200 pages of documentation (Moynihan et al. 2004). Despite its formidable bulk, the consortium and the World Bank worked hard to disseminate the EMP in Chad and outside of it. In contrast to other sources of information the consortium gathered or produced (Leonard 2016; Rosenblum and Maples 2009), the EMP was visible and available. Esso and the World Bank posted the EMP to their project websites, and the World Bank distributed the document through its Public Information Centers in major cities around the world. In Chad, the consortium distributed the EMP in French and English on CDs and delivered print copies to government offices and NGOs. The EMP was also available in special reading rooms the consortium set up in the oil field region and along the route of the pipeline, where it was displayed in long rows of blue binders that filled entire bookshelves. People in the oil field region were not expected to read the EMP; the document was dense and technical and was printed in French and English. It was, however, visible and available, and the sheer bulk and physicality of the EMP ensured that it would be seen.

The consortium said the EMP was the product of extensive environmental, social, and geophysical studies and of consultation with local communities. The volumes of the EMP contained dozens of plans for how the consortium would act in the future—including plans for dealing with oil spills, managing project waste, compensating and resettling farmers who lost land, ensuring road safety, and communicating with residents of the oil field region—but they also described the processes by which those plans had been developed in the past. Transnational companies have argued that corporate social responsibility is an answer to the ethical dilemmas of capitalism, especially in the extractive sector. The EMP was therefore not just a roadmap; it was also a discursive performance of corporate ethics in which global standards and industry best practices played prominent roles.

The volume of the EMP that attracted the most attention among residents of the oil field region was volume 3, also known as the Compensation and Resettlement Plan, which described the consortium's plan for acquiring land for the project. In this Plan, the consortium emphasized its consultation of local communities and its use of a cultural anthropologist to enhance its dialogue with residents of the oil field region (Mallaby 2004). On its website and in its reports, the consortium published photographs of the consultation sessions showing the anthropologist sitting cross-legged among groups of villagers, inspecting their homes, and drawing figures for them in the dirt. The consortium cited her knowledge of the region and her fluency in the local language as assets that aided in the development of compensation and resettlement programs that were sensitive to "local African cultural values" (EEPCI May 1999b; Volume 3, Section 1.7.3). The consortium said that the anthropologist and her assistants held nearly 5500 consultation sessions between 1993 and 2003—"one of the most extensive public consultation efforts for a single project in the history of Africa" (ExxonMobil nd).

The consortium described the consultation sessions as occasions for residents to express their demands, desires, and preferences and to hash out the terms of land transfers and their own compensation and resettlement. In the EMP, the consortium depicted consultation sessions as venues for open and free-wheeling exchanges that lasted "four to five hours" and were ended "by the people attending, when they felt they had adequately expressed their ideas and opinions" (EEPCI May 1999b; Volume 3, Section 2.2.7). The consortium said people could bring up any topic they wanted to discuss in these sessions and that "all groups, including less vocal groups, not just the local power structure, had many opportunities to ask questions and state their ideas." Since the anthropologist and her assistants spoke the local language, the consortium claimed that they "captured for the record most nuances, contentious issues, and informal comments, as well as commendations and recommendations." The compensation and resettlement programs the consortium developed were put forward as products of a process that mimicked the local political process and reflected a thorough vetting of ideas and exhaustive and inclusive deliberations.

Dinah Rajak (2011) argued that these and other types of ethical performances "manufacture a form of consensus which marginalizes alternative visions or critique" (19) while at the same time "mystifying the dynamics of power at work" (10). In fact, these accounts of the public consultation program were difficult to square with my observations or the experiences of the families I followed. Farmers in *canton* Miandoum referred to the anthropologist as the *tête pensante* (the "mastermind") behind the consortium's land expropriation scheme, and they found the community consultation sessions alienating and frustrating. The mood at the sessions I attended was argumentative and confrontational; sessions devolved into shouting matches between angry farmers and the anthropologist or her assistants who were instructed to attend sessions together, keep the meetings small, and leave at the first sign of agitation or unrest (Leonard 2016).

Locals viewed the Compensation and Resettlement Plan as flawed and unfair. They rejected key assumptions underpinning the consortium's land expropriation scheme, including the claims that fallowed land was communal property; that land in Chad belonged to the state and farmers should therefore be paid for their labor but not for the land itself; and that the person responsible for clearing the field in the season of its expropriation should be the beneficiary of the payment, without taking into account the persons recognized as owners of the land under customary tenure arrangements or how the land had been used in the past.

The EMP revealed but it also concealed. In his book on the Baku-Tbilisi-Ceyhan pipeline, Andrew Barry (2013) reprised the work of Georg Simmel (1950) to remind his readers that calls for transparency or more openness do not lead to a parallel decline in the cache or store of secrets. Instead, Simmel argued, when more information is put out in the open, the nature of what is kept secret changes, along with the value of those secrets. While the EMP laid out a detailed, step-wise, and auditable process for land expropriation, one of the things it concealed was the friction surrounding the production and implementation of its component plans. It was not just the EMP; most project documentation obscured these tensions and exceptions were both rare and revealing.

In a document the consortium published to demonstrate its responsiveness to complaints about the project, one of the items the consortium included was how farmers and activists objected to its use of "armed gendarmes" at community consultation sessions (EEPCI 1999a; sec. 9.34). The consortium said it needed armed guards due to "political insecurity" in the region, but that it had responded to the complaint by counseling the guards about "the conduct expected of them during this process" and then, later, by using them only to conduct "reconnaissance" missions prior to the sessions (EEPCI 1999a; sec 9.34). Yet, the document said nothing about how negotiations over land expropriation might have been shaped by the presence of armed guards, nor did it link their use to purges the government was carrying out at the same time of activists and opponents of the project (Amnesty International 1997, 2005).

Transparency required a sprawling apparatus: reading rooms, CDs, blue binders, anthropologists, community consultation sessions, and the EMP. It also required expert monitors whose reliance on the EMP shaped the nature of the information they disclosed to the public. The IAG's trip reports read as mundane updates on the day-to-day operations of the project. In each report, the IAG described dozens of small operational snags in the implementation of the project and gave general management advice on how to resolve them. For instance, following its third monitoring trip to Chad in June of 2002, the IAG wrote:

Cases of mismanagement of individual cash compensation by their recipients have been noted and reported to the IAG.

For upcoming compensations, it would be worth implementing incentive measures to encourage recipients toward savings and investment. Specifically, as in Cameroon, the eligibility threshold for compensations in kind in Chad could be reduced from 70,000FCFA to 30,000FCFA, and the domiciliation of funds with local financial institutions could be encouraged. Extension of the savings and credit structures . . . could also promote the local deposit of savings.

(IAG July 2002, 14)

The monitoring report the IAG submitted following its 12th trip to Chad in April and May of 2007 read, in part:

The biggest complaint heard by the IAG during its visit . . . concerned the time Esso took to process applications for compensation and claims. The group was informed of one impending file dating back to 2004 and of several others that are taking a surprisingly long time to process . . . Esso has room for significant improvement, especially now that it has a much more functional and complete database of claims and compensation, which will enhance its ability to follow up on cases and to standardize its working methods.

(IAG July 2007, 5)

The excerpts reprinted above were chosen randomly and are unexceptional. The issues the IAG drew attention to in these and other reports were those that related directly to the terms and provisions of the EMP. Farmers were not able to effectively manage the cash payments they received, and the process of paying compensation and responding to claims took too long. After describing these bottlenecks, the IAG suggested solutions: allowing farmers to receive compensation in-kind instead of receiving cash, extending the banking system to the oil field region to allow farmers to save money, and speeding up the compensation process and the response to claims. The IAG's observations and suggestions never exceeded the boundaries of the EMP or raised fundamental questions about the project's design. They illustrate Roland Kapferer's observation that "criticism now stands for a set of palliative strategies. It has become a form of tinkering or fine tuning" (Kapferer 2003, 149). The problems the IAG disclosed were problems that could be easily resolved and were amenable to administrative, managerial, or technical fixes. In other reports, the consortium recommended regulating the speed of project vehicles to reduce dust; notifying disgruntled entrepreneurs about the outcomes of their unanswered bids for contracts; devising a plan to inform villages that were no longer eligible for community compensation; and so on.

Monitoring de-politicized the project by containing the kinds of problems and objections that were aired about the project as well as the nature of the remedies that might be sought. The IAG never mentioned, for instance, that before the project started a coalition of NGOs had pressed the World Bank— unsuccessfully—for a moratorium on the project until banking facilities could be extended to the oil field region. Nor did the IAG question the superstructure of the compensation and resettlement program, which was hotly contested by residents throughout the duration of the project.

Oxfam America's evaluation exercise extended the transparency apparatus by adding another layer of monitoring and suggesting that monitoring the monitors was an appropriate role for international NGOs to take up. The knowledge and information required to monitor expert monitors, Oxfam suggested, could be captured by a list of "best practices" (Oxfam America 2011, 8). These are laid out in *Watching the Watchdogs* and include making sure the public is consulted about panel selection; having panel members "report directly to high-level decision makers"; and ensuring that panels have an "independent secretariat" and "independent website" (Oxfam America 2011, 8). These and other best practices—the report lists 29 of them—transform the ethical project of the oil and gas industry into a technical exercise to be managed by layers of monitoring and audit to be carried out with the assistance of checklists, protocols, and manuals. *Watching the Watchdogs* pulls more people, institutions, and resources into the vortex of transparency, illustrating how transparency functions as an instrument of governmentality.

The IAG engaged in a similar exercise on the ground in Chad in instructing the CTNSC, the government monitoring body, and other monitoring bodies in how to do their jobs. The IAG wrote about "the lack of effective monitoring" by the CTNSC (IAG December 2002, 13)—a body that was funded by a World Bank loan to Chad but was not fully staffed or operational for years—and about the need for that body to "perform accurate, thorough and professional verification of each EMP requirement" (IAG June 2003, 4). Andrew Barry argued that "extractive industry transparency is not just intended to make information public, but to govern the constitution of a public that is interested in being informed" (Barry 2013, 73). In Chad, this work began with local monitoring bodies. Whether these bodies would eventually expand their purview and learn to ask other questions, beyond those that relate to compliance with the EMP, as Barry suggests, is an open question. In this project, the actions of the CTNSC were guided and channeled by the IAG, which worked to keep the attention of the CTNSC focused on the terms and provisions of the EMP and not on broader or more expansive questions about the ethics of the project.

The effect of the intense focus on the EMP in the Chad project was to reinforce the idea that global oil companies were committed to the standards regime. The IAG's observations reprised the terms of the EMP, mostly to register the consortium's compliance with them, and it encouraged other monitoring bodies to do the same. The CTNSC's monitoring reports repeat information the consortium provided, reading at times as verbatim copies of the consortium's own reports, and show limited evidence of any parallel data collection or independent monitoring. The CTNSC depended on the consortium to varying degrees over the life of the project to carry out its monitoring tasks. It requested access to the consortium's base camp, work facilities, computers, and electricity and its members asked for the consortium's assistance in securing lodging, meals, and transportation (CTNSC 2001; IAG December 2001; Omeje 2013). The IAG, the CTNSC, and other monitoring bodies posted their reports to their own websites and the World Bank diffused their reports via its website for the project, extending their reach and cementing their claims.

## Wars of words

Occasionally the IAG and other monitoring bodies were critical of the actions of the consortium or accused it of failing to adhere to the EMP. These occasions were rare—which is not surprising, since the consortium wrote the EMP and used the document as its operations manual. But deviations from the EMP illustrated Mol's arguments about the "transformative powers" of environmental information (Mol 2006, 497), and specifically how "informational governance of the environment no longer has simple causalities between a sender of environmental information and a receiver that acts on this information, or between an expert-led monitoring of emissions and environmental qualities, and a state-led environmental action" (Mol 2006, 505). In Chad, the state was sidelined, or was minimally involved in decisions about environmental governance, which emerged out of competing information claims and wars of words. The example I describe in this section of the chapter has to do with a program called the Land Reclamation and Return Program that was proposed by the consortium as a way to mitigate the impact of land loss on families in the oil field region.

The Land Reclamation and Return Program became a focus of expert monitors because not long after oil began flowing through the pipeline, the consortium was confronted with a series of technical challenges that required it to expand the geographic scope of the project and take more land than originally planned. The oil in the Doba Basin fields was heavy and viscous and well pressures fell more rapidly than the consortium anticipated. To boost production, the consortium decided to develop additional satellite fields and drill more wells, but this required the territorial expansion of the project and upended the terms of the EMP and the calculations on which the project's risk mitigation strategies had been based. To increase production and profits, the consortium had to recast certain provisions of the EMP and these efforts put it in conflict with the IAG, other project monitors, and activists, who had taken up the EMP as the "enforcement document" on social and environmental issues.

According to EMP, the consortium was to expropriate 754 hectares of farmland that would be permanently occupied by project infrastructure. The consortium would take more land, but any additional land would be needed only temporarily, to store supplies, stage construction activities, or excavate the earth for sand, soil, or rock. The land taken by the consortium for temporary use would be returned to farmers when it was no longer needed, but as soon as possible to minimize disruption. Yet, when the consortium encountered problems and its land needs ballooned, it was forced to add measures to reduce its footprint in the region. The consortium proposed, for example, to reduce the size of its well pads and installations and it pledged to return temporary-use land to farmers as quickly as possible. Yet, there were chronic delays in the implementation of the Land Reclamation and Return Program and these delays became a recurrent theme in the IAG's monitoring reports.

The IAG first wrote about the consortium's backlog of decommissioned land after its visit to the oil field region in April and May of 2003.[1] In subsequent reports, the IAG reminded the public that the consortium remained noncompliant with the EMP and implored the consortium to make the restitution of unused land a top priority. In report after report the IAG proposed revised deadlines for the consortium to clear its backlog of unused land. In mid-2004, the IAG stressed that efforts to restore and return land "require[d] prompt measures" (IAG July 2004, ii) and urged the consortium to demonstrate a "greater awareness of the rural calendar" (IAG July 2004, 15) by turning over land ahead of the rainy season. Several months later, the IAG characterized the continued lack of progress as an "on-going issue" (IAG December 2004, ii) and a "major delay" (IAG December 2004, 19) and by mid-2005, its language became more formal and pointed. In an extended commentary on the hoarding of unused land, the IAG called the consortium's failure to take action "an urgent environmental issue" that "must be addressed" (IAG July 2005, 4). Monitors wrote that the consortium had provided "no convincing explanation" for the delay and that "a spirit of good corporate citizenship" should motivate the consortium to return unused land quickly (IAG July 2005, 5).[2]

Oxfam America's working theory of transparency would suggest that disclosures such as these would pressure companies to comply with normative protocols like the EMP. But instead of returning the land to farmers, the consortium responded to the IAG's revelations with more information. In fact, the consortium reported itself as non-compliant with the EMP while at the same time suggesting that its infractions were not serious or consequential. The consortium put forward its own system of compliance reporting in which it identified three tiers or categories of non-compliance with its own EMP. The consortium assigned numeric values to each self-reported case of non-compliance to reflect the level of its severity (Moynihan et al. 2004). Cases of level 1 non-compliance were situations that were "not consistent with the EMP" but had "no significant impact to an identified sensitive resource" (EEPCI 2005, 15). Cases of level 2 non-compliance were incidents that "could give rise to a serious impact to an identified sensitive resource" and therefore required "expeditious action" to avoid such an impact, while cases of level 3 non-compliance referenced actual "impacts to an identified sensitive source" and called for "lender-sanctioned remedies" (Moynihan et al. 2004, 10).

The consortium described this reporting system, with its progressive gradations of severity, as an "early warning system" (EEPCI 2005, 15).

Management wrote that it had "empowered" employees to report cases of non-compliance so that it could take corrective action "well before [the cases of non-compliance] become serious enough to cause damage" (EEPCI 2005, 15). By the consortium's account, cases of level 1 and 2 non-compliance referenced problems that would, ideally, spur action. The documents describing this early warning system include color-coded pyramids that transition from yellow at the base to red at the tip and suggest by their tapered shape the success of the consortium in addressing its self-identified operational flaws. Indeed, in the first 3 years of the project the consortium reported 478 cases of level 1 non-compliance, 66 cases of level 2 non-compliance, and only 3 level 3 violations (Moynihan et al. 2004). By this count, the vast majority of self-identified cases of non-compliance with the EMP had no "significant impact" on the oil field region because they were effectively managed before such an impact could materialize.

These controversies over social and environmental risk mitigation strategies were not fought out through empirical studies on the ground. No evaluation was conducted of the actual effects of the consortium's stockpiling of land on crop yields, soil fertility, food security, or hunger among families in the region. As Mol predicted, they relocated to information. The IAG, acting in its role as objective and neutral monitor, repeated the consortium's claims, noting that it had "voluntarily recorded a level 2 non-compliance situation with the EMP in order to stigmatize this negative performance" (IAG November 2005, 7) even while it continued to urge the consortium to return land to farmers and adopt technologies that were less "land-hungry" (IAG November 2006, 5). In visit after visit, year after year, the IAG reported the amount of land the consortium occupied; its failure to return unused land to farmers; and the gap between the EMP and the consortium's actions, yet its reports had no impact on the consortium's practices. The consortium continued to claim that its actions—while not in compliance with the EMP—were part of an effective system of voluntary self-management.

Environmental information was a resource that allowed the consortium to take and hold more land, expand the project, and increase its productivity and profits. Expert monitors were not ancillary to this process; they did not stand outside the process as neutral and objective observers. The consortium responded to the IAG's recommendations by reporting itself as non-compliant, thereby demonstrating rhetorically its commitment to the standards regime while at the same time rewriting the terms and conditions of the EMP and its agreements as a partner on the project. Experts were necessary tools for disseminating this information and building public trust in the process (Giddens 1990), and they were therefore a vital addition to the transparency assemblage in this project. As Mol and others have warned, realizing "the democratic and emancipatory potential of the network society" will be difficult when information is a capital asset and transnational companies have a virtual monopoly, as they did in this project, on the ability to produce, manage, and disseminate it (Mol 2006; Paehlke 2003).

## Conclusion

Expert monitors are new additions to the transparency assemblage in extractive industry projects and are quickly becoming institutionalized as critical elements of that assemblage. The notion that experts are objective, neutral, or independent has been used to argue that their involvement in these projects will exert pressure on the oil and gas industry to operate according to ethical protocols, but the experience of expert monitors in the Chad project suggests that this idea is, indeed, "rather naïve" (Mol 2006, 498) and is itself an informational resource or a capital asset for global oil companies. It is therefore not surprising that companies themselves are advocates of expert monitors. The consortium, the World Bank, and members of the IAG repeatedly referenced the "independence" of the panel, as though repetition might reassure skeptics or make the characterization stick. After all, the IAG was paid by the World Bank and hosted by the consortium when it traveled to Chad—points not lost on people who were used to being reminded about principles of good governance.

Transparency is a technique of governance that shapes and contains public debate about extractive industry projects. Experts who monitor oil and gas projects are thoroughly entangled in networks and flows of information; they do not stand outside those networks and flows and they do not occupy privileged positions from which to make independent or objective assessments. The adoption of expert monitors in oil and gas projects and efforts by NGOs to monitor those monitors mark important expansions of the transparency assemblage but clarify little.

Members of the IAG had the relevant expertise to be critical of the project and of the EMP and its provisions, as well as an open mandate and a sizeable budget. Yet, they defined their mission narrowly, taking up the EMP as a kind of checklist against which to evaluate the ethical conduct of the consortium and the project's progress toward its poverty reduction objectives. Why would they limit themselves to verification of a consortium-authored document?

One answer, though an insufficient one, is pragmatic. The document provided a convenient template for the IAG's semi-annual monitoring trips to the oil field region. In theory, the IAG could explore any leads it chose to pursue, yet in practice it was challenging for the group to move around the region and to interact with residents without the assistance of the consortium. Like the CTNSC, the IAG relied on the consortium for logistical support. Its members flew into the region on the consortium's private airplanes; stayed at its base camp in Komé; and traveled overland in the consortium's vehicles, accompanied by consortium staff. The IAG's reliance on the consortium was so complete that residents of the oil field region were unable to distinguish expert monitors from consortium staff or World Bank personnel. Taking up the EMP as a template or checklist simplified and streamlined the work of the group and made it feasible to monitor the project while spending minimal time in Chad and the oil field region. The EMP also made it possible to monitor the project without needing to spend too much time communicating directly with residents or being bogged down by detailed local knowledge.

The EMP simplified and streamlined the work of the IAG, but the focus on the EMP was about more than getting the job done expeditiously. Roland Kapferer has described consultancy—a term that in his usage encompasses expert monitors—as a "symptom of the post-democratic age in which we live" (Kapferer 2003, 145) and as an "intellectual compromise" with a market driven system based on flexibility and economic efficiency that impels consensus rather than critique (Kapferer 2003, 149). Within this system, experts are reduced to performing a "police function" and to "tinkering" and "fine tuning" (Kapferer 2003, 149)—to making things better at the margins. The IAG's monitoring reports are exemplary exhibits of this. In Chad, the IAG operated entirely within the system of ethical protocols established by the consortium and the World Bank. It never formulated fundamental critiques to the project or proposed alternatives to the system of rational management of oil production ordered by the EMP. George Monbiot (2004) was writing about political leaders, and specifically the former Brazilian president Lula, who was operating under the constraints of the International Monetary Fund, when he said that despite the best intentions of political leaders, "they become technocrats, managers of the conditions thrust upon them." The same might be said of experts, including those attached to the pipeline project in Chad and of Oxfam America's efforts to codify methods for watching the watchdogs that adopt and extend the logic of the World Bank and add another layer of policing.

The effect of expert monitors on shaping the project in Chad was minimal— which is the point of "tinkering" and "fine tuning"—except in one important sense. The consortium's ability to blanket the world with documents, CDs, blue binders, reading rooms, websites, and color-coded pyramids had the effect of constituting the oilfield region as a space of transnational governance based on standards and the consortium as a group of private oil companies committed to corporate social responsibility and a World Bank-approved standards regime. In endorsing the EMP and taking it up as "the enforcement document . . . for the duration of the project" (IAG February 2004, 2), the IAG amplified the visibility of that document and the consortium's commitment to complying with its provisions. By placing the EMP at the center of its monitoring efforts and encouraging other monitoring bodies to do the same, the IAG drew attention to the standards regime and to the consortium's procedural compliance with it, limiting the scope of ethics and the conduct to be investigated and containing the kinds of questions asked and the types of solutions offered.

## Notes

1  This is what the consortium wrote following its site visit:

> The consortium is committed to reclamation of temporarily used soils in time for the next growing season. During the IAG's visit to the oil zone, reclamation operations on arable lands briefly used by the Project had not yet begun due to a

peak in well-drilling operations in the Miandoum Field development zone . . .
Timely soil reclamation is a sensible course of action for correcting impacts on, and
nuisance to, the population. It is important that the Consortium exercise due
diligence in providing for it. The Consortium must institute soil regeneration
operations so that a maximum possible amount of tillable land is returned to growers
in time for the next planting season or, failing that, provide fair compensation. It
must provide information on the issue that is clear and concise so that interested
parties know what to expect.

(IAG June 2003, 9)

2   A fuller version of the IAG's statement follows:

During the Project's construction phase, part of the land in the Oil Field Development
Area (OFDA) and along the pipeline route was annexed permanently and part was
annexed temporarily. Occupation of the land is governed by the Consortium
Agreement and the Petroleum Code, along with the contractual agreement of the
EMP. The area of land occupied and the length of occupation were taken into
account when the Project's compensation plan for area residents was prepared . . .
The IAG again noted serious delays in EEPCI's program to restore land and turn
it over to area residents, thereby depriving them of arable land. No convincing
explanation was forthcoming, other than the possible need for future access. Although
the EMP does not require specific compensation for such delays, a spirit of good
corporate citizenship should motivate Esso to return this land without delay. The
IAG draws the attention of EEPCI and the Government to this issue, which was
raised previously by the External Compliance Monitoring Group (ECMG) but
which still has not been adequately addressed to date. Permanent occupation of
the land with drilling platforms larger than originally planned, and especially more
wells, since Esso is planning to drill up to 100 more wells than stated in the EMP,
reduces the amount of land available for farming and consequently, places additional limits on the population's farming activities . . . In an area where increasing
pressure is being exerted on the land for farming and herding needs, the additional
Project-related pressure is an extremely sensitive issue that needs to be addressed
soon.

(IAG July 2005, 4–5)

# References

Amnesty International. 1997. *Extrajudicial Executions: Fear for Safety*. AFR December
20. London: Author.
Amnesty International. 2005. "Contracting Out of Human Rights: The Chad–
Cameroon Pipeline Project." London: Amnesty International. www.amnesty.org.uk/
files/pol340122005en.pdf.
Auty, Richard M. 1993. *Sustaining Development in Mineral Economies: The Resource Curse
Thesis*. London: Routledge.
Barry, Andrew. 2013. *Material Politics: Disputes along the Pipeline*. Malden, MA: Wiley-
Blackwell.
Castélls, Manuel. 1996. *The Rise of the Network Society*. Volume 1 of the Information
Age: Economy, Society and Culture. Malden, MA: Wiley-Blackwell.
Castells, Manuel. 1997a. *End of Millennium*. Volume 3 of the Information Age:
Economy, Society and Culture. Malden, MA: Wiley-Blackwell.
Castells, Manuel. 1997b. *The Power of Identity*. Volume 2 of the Information Age:
Economy, Society and Culture. Malden, MA: Wiley-Blackwell.

Comité Technique National de Suivi et de Contrôle (CTNSC). 2001. *Projet de Développement Pétrolier, Rapport Trimestriel no. 2, Deuxième Trimestre 2001.* Washington, DC: World Bank.

Dolan, Catherine, and Dinah Rajak. 2011. "Introduction: Ethnographies of Corporate Ethicizing." *Focaal: Journal of Global and Historical Anthropology 60:* 3–8.

Esso Exploration and Production Chad Inc. (EEPCI). 1999a. *Environmental Assessment Executive Summary and Update.* Irving, TX: EEPCI. http://documents.worldbank.org/curated/pt/289441468743146491/pdf/multi-page.pdf.

Esso Exploration and Production Chad Inc. (EEPCI). 1999b. *Environmental Management Plan—Chad portion.* May. Irving, TX: EEPCI.

Esso Exploration and Production Chad Inc. (EEPCI). 2005. *Chad/Cameroon Development Project: Project Update No. 18, Mid-year 2005.* Irving, TX: EEPCI. http://essochad.com/Chad-English/PA/Files/18_allchapters.pdf.

ExxonMobil. nd. Consultation. www.exxonmobil.com/Chad/Library/Photo_Video/web_photoalbum/english/04b.html.

Giddens, Anthony. 1990. *The Consequences of Modernity.* Cambridge: Polity Press.

Haufler, Virginia. 2010. "Disclosure as Governance: The Extractive Industries Transparency Initiative and Resource Management in the Developing World." *Global Environmental Politics* 10 (3): 53–73.

International Advisory Group (IAG). December 2001. *Report of Mission to Cameroon and Chad, November 14–25, 2001.* December 21. Montreal: IAG.

International Advisory Group (IAG). July 2002. *Report of Visit to Chad, June 3–17, 2002.* July 12. Montreal: IAG.

International Advisory Group (IAG). December 2002. *Report of Visit to Cameroon and Chad, October 15–November 4, 2002.* December 11. Montreal: IAG.

International Advisory Group (IAG). June 2003. *Report of Visit to Chad and Cameroon, April 21–May 10, 2003.* June 18. Montreal: IAG.

International Advisory Group (IAG). February 2004. *Report of Visit to Chad, December 5–21, 2003.* February 12. Montreal: IAG.

International Advisory Group (IAG). July 2004. *Report of Visit to Chad and Cameroon, May 17–June 5, 2004.* July 9. Montreal: IAG.

International Advisory Group (IAG). December 2004. *Report of Mission 8 to Chad, October 10–16, 2004.* December 3. Montreal: IAG.

International Advisory Group (IAG). July 2005. *Report of Mission 9 to Chad and Cameroon, May 15–June 6, 2005.* July 11. Montreal: IAG.

International Advisory Group (IAG). November 2005. *Report of Mission 10 to Chad and Cameroon, September 25–October 18, 2005.* November 24. Montreal: IAG.

International Advisory Group (IAG). November 2006. *Report of Mission 11 to Chad, September 24–October 14, 2006.* November 8. Montreal: IAG.

International Advisory Group (IAG). July 2007. *Report of Mission 12 to Chad and Cameroon, April 30–May 24, 2007.* July 18. Montreal: IAG.

Kapferer, Roland. 2003. "It's a Small World After All, or, Consultancy and the Disneyfication of Thought." *Social Analysis* 47 (1): 145–151.

Karl, Terry Lynn. 1997. *Paradox of Plenty: Oil Booms and Petro States.* Berkeley: University of California Press.

Leonard, Lori. 2016. *Life in the Time of Oil: A Pipeline and Poverty in Chad.* Bloomington: Indiana University Press.

Mallaby, Sebastian. 2004. *The World's Banker: A Story of Failed States, Financial Crises, and the Wealth and Poverty of Nations.* New York: Penguin Books.

Mol, Arthur P.J. 2006. "Environmental Governance in the Information Age: The Emergence of Informational Governance." *Environment and Planning C: Government and Policy 24* (4): 497–514.

Mol, Arthur P.J. 2008. *Environmental Reform in the Information Age: The Contours of Informational Governance.* Cambridge: Cambridge University Press.

Mol, Arthur P.J. 2010. "The Future of Transparency: Power, Pitfalls and Promises." *Global Environmental Politics 10* (3): 132–143.

Monbiot, George. 2004. *The Age of Consent.* Lecture to the Royal Society of Arts. London: RSA.

Moynihan, Kelly J., Clayton F. Kaul, Ed R. Caldwell, Ulrich L. Sellier, Neil A. Daetwyler, Gary L. Hayward, and Joey V. Tucker. 2004. *Chad Export Project: Environmental Management and Monitoring Process Systems.* SPE 86721. Richardson, TX: Society of Petroleum Engineers.

Omeje, Kenneth C. 2013. *Extractive Economies and Conflicts in the Global South: Multi-Regional Perspectives on Rentier Politics.* London: Ashgate.

Oxfam America. 2011. *Watching the Watchdogs: Evaluating Independent Expert Panels that Monitor Large-scale Oil and Gas Pipeline Projects.* Washington, DC: Oxfam.

Paehlke, Robert C. 2003. *Democracy's Dilemma: Environment, Social Equity and the Global Economy.* Cambridge, MA: MIT Press.

Polgreen, Lydia. 2008. "World Bank Ends Effort to Help Chad Ease Poverty." *New York Times,* September 10.

Rajak, Dinah. 2011. "Theatres of Virtue: Collaboration, Consensus, and the Social Life of Corporate Social Responsibility." *Focaal: Journal of Global and Historical Anthropology* (60): 9–20.

Rosenblum, Peter and Susan Maples. 2009. *Contracts Confidential: Ending Secret Deals in the Extractive Industries.* New York: Revenue Watch Institute.

Simmel, Georg. 1950. "Secrecy." In *The Sociology of Georg Simmel,* translated and edited by Kurt H. Wolff, 330–344. Glencoe, IL: The Free Press.

World Bank. nd. International Advisory Group. http://web.worldbank.org/archive/website01210/WEB/0__CO-36.HTM.

World Bank. 2001. "World Bank Appoints International Advisory group on the Chad-Cameroon Petroleum Development and Pipeline Project." February 21. News release no. 2001/235/S. New York: World Bank. www.ciel.org/Publications/Chad Camadvisorygroup.pdf.

# 10 Post-script

## Mapping neo-extractive frontiers across Africa and Latin America

*Brenda Chalfin*

The challenge of this volume lies in its breadth and brevity. The governance of extractive economies is an enduring concern in a world industrial epoch built on the exploitation of mineral reserves and their conversion into energy and other essential inputs. Despite differences between the founding features and drivers of industrialization and the material conditions and contradic-tions of late-modernity, extractive regimes continue to flourish. As this collection of chapters demonstrate, the opening of new resource frontiers continues unabated—from the deep seas of São Tomé and Príncipe, to Chad's massive subterranean rifts and the methane-rich volcanic lakes of Central Africa. However, in contrast to the continuities of long ascendant mineral and hydrocarbon economies (oil, natural gas, gold, copper, and tin) and associated strategies of accumulation, in the opening decades of the new millennium there is less certitude regarding the conventions of extractive governance.

The twenty-first century is for sure an incompletely charted political field. The political blocs and hierarchies that marked the twentieth are fundamentally unsettled. The nation-state, and most certainly, the developmental state, is no longer in clear ascendance either in the global north or global south. While national states have not disappeared, they increasingly align and combine with transnational capital and international regulatory fields (Chalfin 2010; Sassen 2006). With the dissolution of states' role as arbiters of social norms and social protections, non-governmental and humanitarian organizations flourish, buoyed by the growing span of transnational cultural projects (Adunbi and Chalfin 2017).

Generating a highly variegated and decidedly *lateralized* political field, the upshot of this reordering is that although extraction is here to stay, and will most likely expand apace, the governing frameworks that oversee and enable extractive endeavors are difficult to fully decipher. Focused on eight cases drawn from across Africa and Latin America and key international forums, a systematic probing of this line of inquiry is the core purpose of the volume.

Taken together, the foregoing analyses underwrite three central propositions regarding the character of extractive governance. One, in the current era extraction is generative of a huge governmental apparatus. Reorganizing estab-lished hierarchies, the result is a surplus of governing bodies, procedures and norms, many technical in nature and reliant on the production and circulation of expert knowledge. Two, extractive governance is closely aligned with the logics and prerogatives of environmental governance. Although the two are not identical, they collectively promote, partake of and normalize the language of emergency and crisis. These optics simultaneously invite social activism and tactics of repression. Finally, extractive endeavors, due to their broad scope and intensity, actively reorder the social contract. Altering the nature and expectations of public goods, they reframe the political compact, whether through the acceptance of "public bads" by those effected by extraction, or the tacit recognition of counter-movements by ruling authorities. Contributing to the wider syndrome of governmental expansion around extractive operations, such processes enlarge the terms of public participation and public oversight. Yet, as the preceding chapters substantiate, the hyper-development of extractive enterprise and extractive rule continues to confront that which is ungovernable.

The case studies in this volume forcefully suggest that frameworks and logics operationalized around the governance of mineral, gas, and hydrocarbon extraction are simultaneously symptomatic of and have a formative impact on wider transformations of what is variously called the late, post, or advanced neoliberal state. Though a small sample of the numerous national economies invested in extraction, the examples covered in this book are representative of late-neoliberal political economic conditions across Latin America and Africa. Bolivia, Ecuador, and Peru all self-consciously retooled their extractive industries to reclaim and redistribute rents on the domestic front. Challenging the privileged position of transnational capital, these projects are intertwined in different ways with state rebuilding: whether Bolivia's call for a pluri-national polity, Ecuador's cultivation of repressive capacity, or Peru's protection of strategic international partnerships.

The African cases equally span the spectrum of late-neoliberal arrangements. In each, despite the embrace of different sorts of economic extroversion, state actors utilize extractive economies first and foremost for political survival. From Rwanda's lure of foreign direct investment and international aid through the guarantees of entrenched authoritarianism, to Chad's courting of development capital as a means to evade popular and international accountability, to São Tomé and Príncipe's banking on economic speculation as the path to growth and international recognition or the Nigerian state's quest for market share at any price. Indeed, these overlapping dynamics suggest that the concerted move toward neo-extractivism in Latin America noted by Perreault among others (Burchardt and Dietz 2014) may well be at work in Africa, even if they have not been self-consciously theorized as such. To paraphrase Perreault, they include the intensification of resource extraction in the context of favorable global markets, state efforts to capture a greater share of the rents, and a nominal redistribution of those rents through various social programs and investments. A postscript to the foregoing case studies and a prelude to a much longer conversation, the remarks below address the common patterning and contradictions of governance across these different national and regional projects.

## The governmentalization of extraction

It is clear that the advance of the extractivist project in Africa, Latin America and more broadly is closely tied to the proliferation of governance bodies, governmental functions and governing mandates. Extending governance into new domains to generate a vast network of institutions devoted to rule-making, oversight, and enforcement in the name of public, national, or other collective interest, these arrangements carry their own structural contradictions and impediments even as they render governance a legitimate and well-recognized wing of the extractive sector.

Pulver's account of the inner-workings of environmental non-governmental organizations (ENGOs) pitched to oil industry representatives at the 2015 UN climate conference vividly illustrates this process of "extractive governmentalization." Expanding the number and ambit of ENGOs on the international front, this move is generative of "governance" but not necessarily generative of government or governing. Rather, the forms of transnational environmental advocacy captured by Pulver serve as an "auxiliarization" of governance promoting popular alliances and spectacle to push for a more open and inclusive deliberative realm. In this popularization of governance, the divide between the political inside and political outside is blurred. With rising numbers and types of non-state observer organizations operating in these international forums, the quest for an unbounded ambit of global governance comes into view. Though occurring on a global stage and scale, this process is also at work on the national level, as Weszkalnys's portrait of São Tomé and Príncipe's courtship of Jeff Sachs reveals. Cutting through the usual hierarchies of governance by drawing on the knowledge-based authority of academic experts in this West African micro-state, a technocratic rhetoric of energy and environmental politics becomes part of public discourse and consciousness.

In terms both institutional and ideological, Leonard's and Grovogui's detailed portrayals of Chad's Doba Basin oil pipeline project illustrates the process and pitfalls of governmental propagation surrounding extractive endeavors. A mega-project leveraging substantial international investment to jumpstart extraction-for-development, Leonard recounts, international financial institutions and other funders drive the Chad-initiative mandating project oversight on fronts fiscal, environmental, and social. Generating an alphabet soup of entities devoted to monitoring and evaluation (M&E)—CCSRP, CTNSC, ECMG, IAG, and more—the result is what Leonard calls a "sprawling apparatus of transparency." Designed to "watch the watchdogs," to borrow Oxfam's phrase, this cumbersome administrative superstructure is ultimately ineffective. All checks without the counterweight of balances or national buy-in, as a form of lateralized rule it is thin on accountability though thick with evidence.

Grovogui's focused investigation of the legal apparatus that grew alongside Chad's oil for development plan identifies the central contradiction of what he calls "advanced neo-liberal legal systems" that, like the monitoring bodies above, mimic preconceived codes and generic best practices. Here law proliferates and carries ideological potency to bestow legitimacy on the Chadian state yet is detached from capacities for enforcement or any meaningful connection to the legal subjects and objects it is deemed to protect—namely, rural property rights. Signaling a key concern running across all the book's chapters, Grovogui looks to the implications for state authority, finding the combined extensification of the rule of law and the intensification of its application, no matter how ill-fitting, a means to consolidate state power in the face of the expansion of extra-state interests and institutions. Compared to the coordinated administrative agendas of M&E, the legal apparatus Grovogui describes, despite its technical limitations, is a much more deeply networked form of power.

Perreault's examination of the Bolivian case also reveals the centrality of law to states' navigation of the often contending interests of extraction-affected communities and international capital. While the populist, "pluri-national" pretensions of Bolivia's President Morales differ from the unabashed authoritarianism of Chad's ruling regime, in both settings extraction boosts legal production. Perreault considers the co-production of legal provisions that explicitly privilege the mining sector and those that bestow rights and recognition to indigenous values and communities in the service of informed consultation and notions of "living well." Like Grovogui, Perreault notes the symbolic work of law-making to confer legitimacy on already privileged ruling groups. However, he points to an additional register of significance in that these laws, though logically at odds, provide a language for making groups and interests politically and discursively "legible" within the nation at large. Whereas governmental proliferation from Leonard's perspective generates a paralyzing redundancy, the *pluri-legal* Bolivian case suggests looking beyond their legal–rational in/efficiencies to the collective values they confer.

### Environment, exception, emergency

As analyzed by Billo, the legal field surrounding open pit copper mining in the Ecuador/Chile border region tells a very different story. While it too, like that of Chad and Bolivia, bolsters state authority, in the Ecuadorian case the rule of law is backed by brute force. The conditions described by Billo demonstrate how large scale extractive endeavors generate and draw upon "states of emergency" to consolidate operational capacity and undermine opposition. Law is nothing short of a cover for the untrammeled violence of the state. In this case the state mining company, ENAMI, works hand in hand with police and other security agencies to crack down on local and international activists and NGOs. Here we see a diversity of para-political organizations at the margins of the state—international financial institutions, transnational corporations, international non-governmental organizations—as well as state-based entities push the limits of rule. Equating environmentalists with criminals, suspected parties are subject to physical and emotional abuse and a whole range of "para-legal" punishments such as surveillance and detention. Vividly illustrating Giorgio Agamben's (1998) thesis, mining law in the Ecuadorian setting enforces a "state of exception" at once accentuating sovereign prerogative and reducing extraction's detractors to a condition of "bare life."

States of emergency also play a central role in Rwanda's emerging extractive sector discussed by Doughty. Centered on Lake Kivu's large methane reserves formed by underwater volcanic eruptions, Doughty shows how the rapid rollout of this energy production scheme is presented to the Rwandan people and the international community as a necessary exercise in emergency management. Endowing Rwandan's highly authoritarian state with new powers under the aegis of environmental and human disaster prevention, extraction—and with it, the state—is portrayed as an engine of national salvation. In short, environmental emergency justifies the government's exceptional powers. As Doughty perceptively observes, the Lake Kivu Energy Project is embedded in a multifaceted "conversion narrative," transforming dangerous gas to productive energy, natural disaster to political benevolence, scientific fact to popular imaginary. In post-genocide Rwanda, these tropes are far from innocent. By posing nature as a source of violence that can be tamed by human intervention, the types of violence humans visit upon one another fall out of view.

Recounted in its many-layered complexity by Li, the saga of gold-mining induced mercury poisoning in Peru also partakes of tropes of emergency in which understandings of accountability are similarly unfixed. Years after the Choropampa mercury spill, residents still seek redress for bodily and environmental harms. In an economic and medical domain otherwise saturated with information and expert knowledges, Li's interlocutors note the failures of "public reason" as well as broader "epistemic failures" to make sense of personal and communal calamity. Nevertheless, these gaps and denials re-render the social contract, turning insult into opportunity. Spill victims find themselves positioned in wider resource circuits, receiving funds for education, healthcare, and offspring. In this scenario, it becomes clear that despite the foreclosure of "extractive citizenship" built on rights and representation, the position of "extractive subject" based on the recognition of chronic vulnerability and qualification for redress remains a singular entitlement born out of emergency. At the same time, turning emergency into historical consciousness, extractive events and accidents become points of historical reference in their own right, impacting understandings of place and futures for both the principals of extraction and their subjects.

## Social contract, social compromise, and speculative governance

Redrawing the social contract, the kinds of social compromise recounted by Li in the case of Peru can also be found in Africa-based extractive regimes described in this volume. Weszkalnys hones in on the plight of the island nation of São Tomé and Príncipe, where the exploitation of oil reserves has long been anticipated but has yet to materialize. Yet another example of extraction driven governmentalization, the expectation of oil is oddly productive in its own right, spawning all sorts of regulatory, promotional and prospecting bodies. Weszkalnys provides a nuanced analysis of how the extended if not suspended stage of speculation provides São Tomé and Príncipe a window to contribute to the making of oil expertise. The idea of the extractive economy—even without the material markers of actual extraction—serves as national political economic platform and basis for social consensus. Flipping social goods and social bads, this platform of nation-making grows out of and further cultivates the specter of the resource curse—yet another disaster scenario driving extractive projects. A foundation for a collective imaginary as well as the buy-in of international experts and advisors unmoored from formal governmental institutions, STP's social contract becomes displaced on a surprisingly labile and mobile economic idea. All the while, the para-governing capacities of academic experts are emplaced within a developing micro-state.

Adunbi's account of rogue artisanal refineries aside the creeks of Nigeria's oil rich Niger Delta reveals a different set of social and political compromises in the offing. Like the other extractive domains described in this volume, they too partake of a profusion of governmental forms and agents. Adundbi, for instance, refers to the interplay of invisible patrons, direct governors, and community partners in bringing these extractive formations to life. They are based not on speculation and remote figurations but on the materially dense and often dangerous outputs of the Delta's subterranean hydrocarbon stores, brought to life in the labors, lands, and bodies of Delta residents. Officially classified as illegal crude oil bunkering and processing plants, despite the "public bad" of their associated pollutants, these installations are public goods of a sort: providing income and employment, cheap and copious supplies of kerosene to neighboring communities, as well as a potent symbol of resistance to international oil companies and state predation. Given their growing scale and technical success, the rogue refineries present a conundrum to the state authorities whom they have consistently sought to challenge and evade. In this arrangement, we find one of the few examples of a neo-extractivism not driven by the joint interests of state actors and multinational capital. Rather, in play is state recognition, if not endorsement, of a rogue privatization, the spoils of which they do not directly or decisively share, yet which promote social redistribution. The terms of this social contract, Adunbi readily admits, remain unsettled, confounding the boundary between the ungoverned and the ungovernable as new forms of state evasion and collusion emerge out of the never-ending possibilities of technical and political reinvention.

## References

Adunbi, Omolade, and Brenda Chalfin. 2017. "Governance." In *Critical Terms in African Studies*, edited by Adeline Masquelier, and Gaurav Desai. Chicago: University of Chicago Press.

Agamben, Giorgio. 1998. *Homo Sacer: Sovereign Power and Bare Life*. Stanford: Stanford University Press.

Burchardt, Hans-Jürgen, and Kristina Dietz. 2014. (Neo-) Extractivism: A New Challenge for Development Theory from Latin America. *Third World Quarterly 35* (3): 468–486.

Chalfin, Brenda. 2010. *Neoliberal Frontiers: An Ethnography of Sovereignty in West Africa.* Chicago: University of Chicago Press.

Sassen, Saskia. 2006. *Territory, Authority, Rights: From Medieval to Global Assemblages.* Princeton: Princeton University Press.

# Index